T0324115

COLLABORATIVE INTERNET OF THINGS (C-IOT)

COLLABORATIVE INTERNET OF THINGS (C-IOT)

FOR FUTURE SMART CONNECTED LIFE AND BUSINESS

Fawzi Behmann

Kwok Wu

This edition first published 2015

Registered office
John Wiley & Sons Ltd, The Atrium, Southern Gate, Chichester, West Sussex, PO19 8SQ, United Kingdom

For details of our global editorial offices, for customer services and for information about how to apply for permission to reuse the copyright material in this book please see our website at www.wiley.com.

Reprinted with Corrections August 2015

Library of Congress Cataloging-in-Publication Data applied for

ISBN: 9781118913741

Typeset in 11/13pt TimesLTStd by Laserwords Private Limited, Chennai, India

1 2015

Contents

Foreword

I recall sitting in the Bell Labs auditorium in Holmdel, New Jersey in 1980 listening to a lecture by the head of research, Arno Penzias. He had recently won the Nobel Prize for his radio astronomy work on the origins of the universe. But this day he was evangelizing a technology we all knew, but did not fully appreciate; at least that was Dr. Penzias' message. As engineers and scientists we did not lack appreciation for the *invention*, the microprocessor. The Intel 8080 and Motorola 6800 replaced thousands of discrete small- and medium-scale integrated circuits used in random logic designs, cutting development time and costs dramatically. Some in the room were even determined to invent the next generation of microprocessors. That was not what Dr. Penzias was proselytizing. His was a challenge of *innovation* with the microprocessor. That vision was one of thousands of microprocessors in homes, automobiles, and offices. By sharing his vision, he was spurring us to think not about problem solving but about possibilities. At the time, I dare say most of us failed to grasp the full import of his message because of our linear thinking. Engineers in particular are prone to linear thinking. We are skilled at wrapping our minds around a single complex problem, going ever deeper, searching for clever ways to overcome natural barriers to achieve a novel and useful design.

Even in hindsight we think linearly, as does much of the general public. If you ask "who changed America's homes by lighting them with electricity?" you are likely to hear the name of Thomas Edison, not Nikola Tesla and certainly not James Watt. Yes, Edison is credited with the invention of the incandescent light bulb, but he waged a long legal and publicity war against Tesla's invention of alternating current (AC), advocating his own choice of direct current (DC). In that way he impeded, not hastened the lighting of America's homes. Without AC generation and transmission, we would all need DC power plants in our basements to light our homes. Tesla made centralized power a practical and commercial reality. As for James Watt, it was his steam turbines that converted fossil fuel, primarily coal, into electricity so that it could be transmitted over Tesla's network to power Edison's bulb. Without Watt's steam engine, Edison and Tesla's inventions would have been subjects for demonstration in undergraduate physics.

Behind the invention of the Internet is a similar story. Vint Cerf and Robert Kahn are the names that come to many engineers when asked about the origins of the Internet. That recognition is richly deserved. Their contribution of TCP is foundational to today's network, and without their guidance and advocacy, the Internet would not have evolved to the one we know today. Yet most of the general public knows little or nothing of TCP. It is more likely that when they think of the Internet, they see the World Wide Web and the web browser, invented by Tim Berners-Lee and Marc Andreesen, respectively. While the public is not likely to recognize those names either, they do recognize the names of Steve Jobs and Bill Gates, and so they may receive popular credit. That too is deserved, without the personal computer, the Internet would look very different indeed. Perhaps, the most underappreciated Internet engineers are Dr. Emmanuel Desurvire and Dr. David N. Payne. Without their invention of the Erbium Doped Fiber Amplifier (EDFA) or the invention of modern fiber-optic cable, Internet backbones would be operating over coaxial cable at speeds of megabits per second, not terabits per second. We would be stuck with our 38 kbps voice band modems. Remember how much fun it was downloading a song or an image with that? To be fair, it took all of these inventions and contributions to make our Internet a reality.

These stories are tales of the convergence of inventions at a single place and time in history that unleashed floods of innovation that flowed for decades, transforming societies, businesses, and even cultures. No one sat down and decided that to have a successful Internet, they would need a reliable transport protocol, fiber-optic communication, a multimedia web protocol, personal computers, and a browser. However, when they all came together in the early 1990s, innovators and venture capitalists began to see and explore the possibilities. In the late 1990s and early 2000s, the public, investors, and media saw new sites and businesses announced daily, as eCommerce, eBusiness, and B2B (business-to-business) were added to the lexicons of the world. Some companies rose from nothing to great heights, such as Webvan, only to fall again as unsustainable. Others, such as Egghead, saw their entire business model turned on its head. An industry it had helped create, the personal computer industry, and a technology they made popular, the voice band modem, turned on them. Their success made their brick and mortar franchise obsolete. Software could be downloaded without going to the local Egghead store, and much of the software was free.

But all of that is history, where are we now and where are we going? If we have learned anything from the history of invention, innovation, and adoption in the last 200 years, it should be humility. If you need evidence, look at the aftermath of the dot-com bubble or read a 10-year-old article from the popular press on the future of technology. As the President of AT&T Labs, I was often asked about the next great innovations and what they meant for our networks, which ones would drive bandwidth growth in our homes and businesses, and how they would change the way we work and recreate. My answer was that I did not know what applications and innovations would drive our networks and lives; that response never failed to disappoint the interlocutor. But what I did share with audiences inside and outside AT&T were technologies and trends that would shape that future. Those technologies and network trends were the subjects of investigation and innovation of the Members of Technical Staff at AT&T Labs, and I had the great privilege of seeing their work daily. What I could predict was the exponential growth in bandwidth, at 40–50% each year for decades, as postulated at Nielsen's Law, a corollary to Moore's Law, and what I could see were technologies that mattered in the shaping of our future.

In a broad sense, "Collaborative Internet of Things (C-IoT) for Future Smart Connected Life and Business" by Fawzi Behmann and Kwok Wu presents the reader with such a view of emerging technologies, and how at this point in time, they will work together to usher forth another flood of innovations, changing our lives. The theme of the book, Internet of Things or IoT is a term meant to capture the pervasiveness of the Internet, the wide adoption of mobile computing and connectivity, and their incorporation into everyday things in our lives. Those technologies are leading directly to the ubiquity of embedded computing in the most common place and also into the most complex items in our lives. There are already shoes, pet collars, and light bulbs connected to the Internet, and our homes and automobiles are living out Dr. Penzias' vision of thousands of embedded processors. Add to those technologies cloud computing, the introduction of IPv6, and the emergence of Big Data analytics and we begin to see the possibilities and models for adoption that are explored in this book.

Few if any of us can predict the next Facebook or iPhone. But by identifying technologies that matter and a likely framework for their evolution and adoption, we can begin to see the possibilities, much as Dr. Penzias urged us to do back in Holmdel NJ.

G. Keith Cambron
President and CEO of AT&T Labs, retired

About the Authors

Fawzi Behmann

President, TelNet Management Consulting, Inc.

Fawzi Behmann is the president of TelNet Management Consulting, Inc., a results-driven firm incorporated in 2009 in Texas, USA. The focus is in providing consulting services in the areas of empowering smart communications and networking and providing insights in several vertical markets. Company capabilities include the development of global strategic initiatives, products, solutions, training, and support in the areas of Internet of Things (IoT), wireless, public safety, enterprise, industrial, medical, supply-chain infrastructure, and big data analytics.

Fawzi Behmann has many years of experience in global communications and networking spanning supply chain from semiconductor, networking equipment, and service providers in Canada and the United States. This in turn helped in understanding customers' requirements, market and technological trends, developing strategy, and plan of execution applying best of practices.

- With coauthoring the book Collaborative Internet of Things (C-IoT) for Future Smart Connected Life and Business, Fawzi Behmann has pioneered the development of early-IoT system for telecom in the late 1980s and early 1990s based on ITU TMN M.3000 standards.
- As a Consultant/Executive Marketing Director since 2009, he provided support to Power.org, a nonprofit worldwide Trade Association, in advancing $5 Billion dollars Power Architecture (Power PC) processor technology and promoting ecosystem solutions in select key markets. Fawzi develops corporate strategic plans and facilitates business collaboration with developers, academia, and other forums. Key served markets include cellular LTE/Wi-Fi communications and networking and server/big data analytics.
- As a consultant, Fawzi has been supporting public safety projects based on risk management approach. The focus is in the areas of communications and networking for emergency command and control, radio, data networking, and

video surveillance. Fawzi also collaborates with other consortium members and suppliers in defining turnkey integrated solutions-based geographic information system/global positioning system GIS/GPS and following international standards for Fire Fighter Cover Safety Plan.
- As a senior member of IEEE and the Chair of IEEE Communications and Signal Processing Joint chapters in Austin, TX, Fawzi organizes monthly technical seminars, workshop, and outreach programs for the local professionals and academia. He serves as Central Texas PACE chair, NA Distinguished Lecturer/Speaker coordinator, and the Chair of Local Arrangement and Marketing Chair for IEEE International Globecom and leads the automation of PACE program for IEEE USA.

Among other key achievements:

- As a Director of strategic marketing with Motorola/Freescale in the United States, Fawzi developed wireless technology positioning and market trends for products and solutions. He articulated value proposition in supporting scalable broadband traffic, multicore, multi-threading SoC – System on a chip, scalable input/output, and scalable security for diverse markets. Fawzi led networking working group at International Technology Roadmap for Semiconductors (ITRS) in defining networking platform vision and roadmap for the next 15 years, which was issued as a part of ITRS publication.
- As a senior product and solution manager with Nortel Networks, Fawzi defined Intelligent building structured wiring, Internet-based LAN – local area network product management, IP – Broadband services node switch/router product for the edge of the network, and product release of Core WAN – wide area network switch. He supported pilot project serving 10 000 clients for a residential broadband services.
- As a project and team leader and acting section manager, Fawzi was responsible for defining multi-year, $50 million strategic corporate R&D and Network Management program for Teleglobe, Canada (now TATA Communications). He championed the definition, specification, and development of monitoring, control, and supervisory network management system. The system was implemented for Teleglobe Telecom at local, regional, and national levels. Fawzi led the development of state-of-the-art network control center, which was equipped with graphical real-time display, and LED (light-emitting diode) of world-map identifying facility failure and impact on traffic and services.

Fawzi has been an agent of change at three fronts: Moving from analog world to the digital world in the service provider space, penetrating the enterprise space with rapid acceleration of technology to IP, and embedding intelligence into the semiconductor space.

Fawzi has been active in international forums. As a member of the Canadian Delegation team, Fawzi participated in the development of ITU M.3000 standards for Telecom Management Networks in Geneva.

Fawzi organized over 1000 h of technical sessions and international conferences and held over 250 h of media briefings. He has written and published several white papers and has been a keynote speaker and presenter at several conferences domestically and internationally.

Fawzi holds a Bachelor of Science Honors in Mathematics with Distinction from Concordia University, Masters in Computer Science from Waterloo University, and an Executive MBA from Queens University in Canada. Fawzi was a recipient of the Freescale CEO Diamond Chip Award (2008) and recently an IEEE R5 Outstanding Member Award (2013).

Fawzi Behmann can be reached at Fawzi.behmann@telnetmanagement.com.

Kwok Wu, PhD

Head of Embedded Software and Systems Solutions, Freescale Semiconductor.

Dr. Kwok Wu, an award-winning industry veteran and sought-after speaker, has been awarded 2012 Innovator of the Year by ECD – Embedded Computing Design Magazine for his platform approach to Wireless Smart IOT Gateways.[1]

In addition, Kwok was awarded the 2011 Innovative Networking Product Award, from the Broadband World Forum with Secured Broadband multi-service Gateway. He was also a recipient of the 2012 Best Networking and Communication Product Award, Smart Metering at Australia, and New Zealand Summit.

Dr. Wu has many years of diverse experience in advanced embedded systems and software. He has delivered high-performance scalable software platforms and products for Freescale's Power Architecture, ARM, and ZigBee SoCs in the wireless broadband networking, telecommunications, enterprise, consumer, automotive, industrial, smart energy, and health segments.

He has held various executive management positions at AT&T Bell Laboratories, Lucent Technology, Actel, AMD, Lattice, and Freescale Semiconductor. Kwok is a member of the IEEE Computer Society and holds a Treasurer position at the Austin Chapter of IEEE Communications Society, and he holds a PhD, EECS (Computer Engineering) from the University of Texas at Austin.

Kwok Wu, PhD Kwok.Wu@Freescale.com
wu.kwok@gmail.com
Mobile: 1-512-971-5364

[1] http://embedded-computing.com/articles/2012-solutions-freescale-semiconductor/

Preface

Every day, the market is bombarded with information and news about Internet of Things (IoT). This comes in a variety of forms such as articles, books, seminars, and conferences.

This book deals with the explosion of information on IoT with a simplified visionary approach for the future of IoT. Today, we witness a discrete IoT solution (point solution) within a given vertical market. The focus in the future is a collaborative intelligence that will impact our connected life and businesses. The future or next-generation IoT will be called Collaborative Internet of Things (C-IoT) in this book. The focus is on core concept that has impact on improving the quality of our lives and also improving business efficiency.

This book introduces a simple innovative model for C-IoT and a new way of looking at the market. The C-IoT model, in its simplest form, consists of sensing, gateway, and services. Sensing will tap into what matters, and gateway will add intelligence and connectivity for action to be taken at the local level and/or communicate information to the cloud level. The services will capture information and digest, analyze, and develop insights of ways that help improve quality of lives or improve business operation. Relevant standards and technology enablers will be highlighted for each segment of the model. The model will address both present and future IoT opportunities and provide the reader with clear positioning as to where radio frequency identification (RFID), machine-to-machine (M-M), and others fit in the model.

In addition, this book introduces simplified market segmentation for C-IoT using domains and business applications. The three C-IoT domains are 3Is: Individual, Industrial, and Infrastructure. Individual C-IoT represents smart living covering consumer electronics and wearable devices, smart homes, and smart connected cars. Industry C-IoT is for business efficiency, which covers several markets associated with industry such as smart factory, smart buildings, smart machine, and smart retails. Infrastructure C-IoT represents smart communities and cities for sustainable environment and living, which include public transportation and highways, public safety, disaster management, smart education, and smart health care. Business applications such as Health & Fitness can easily span the three domains: Individual (e.g., wearable devices), Industry (e.g., physicians, labs, and hospitals), and Infrastructure (e.g., FDA, law, and enforcement). Collaborative IoT solutions will impact breaking down the barriers between traditional vertical markets and supply chains as the Internet broke down the geographical barriers.

Gaining insights in each IoT domains will result in driving better results in each business application in that domain. Spanning to other IoT domains creates value for a better strategic decision. Take for example, a smart grid of the future; it is feeding into a network of power distribution connecting cities, businesses, and residential buildings. Having a smart meter in the home not only will help to lower the operating energy costs but also will link to the grid and becomes aware of the environment and opportunities that may result in incremental savings. Similar example can be applied to health care from wearable devices for fitness and health monitoring, connecting to physicians and hospitals for diagnostics and treatments to networking with insurance and government agencies for policy and governance. Another example would be video surveillance for homes, enterprises, and public safety. This book highlights several other use cases including tracking and monitoring using RFID, wireless WiFi, 3G for location tracking, GPS, and so on.

Thus, the C-IoT for the future is a disruptive technology that spans all vertical markets causing convergence and breaking down the barriers.

The core of this book is to highlight a series of cases spanning from the requirements to the solution for present market and business opportunities and to explore future opportunities.

The C-IoT enables convergence of several technologies and consequently impacts the overall architecture of the network. We envision a common software platform for most of the vertical markets, adaptation, and customization through applications and special devices to address the specific needs of a given market. This unified smart C-IoT software platform enables one to build and deploy smart C-IoT product, systems solutions, and services for different vertical markets in a quick time-to-market fashion. This will be described in a later chapter of this book including the 4A's and 4S's that characterize smart C-IoT products that would facilitate delivery of Internet of Service (IOS). The 4A's stand for Automated Remote Provisioning and Management, Augmented Reality, Awareness of Context and Location, Analyze, and Take Action, Automate and Autonomous, and Anticipate. The 4S's stand for Simplicity, Security, Smart, and Scalable. Chapter 3 ends with covering the Secured IoT.

This book provides examples of do-it-yourself (DIY) kits aimed at bringing the concept, approach to hands-on experience that inspires innovative thinking in exploring untapped opportunities that improves the quality of humanity and business efficiency in general.

Finally, this book will address the emerging new wave of new devices such as wearable/mobile and cloud technology (local, public, and inter-cloud), analytics, and social media as key building blocks of collaborative IoT distributed intelligence. This book also examines the Collaborative IoT impact on our digital lives and businesses and some of the future challenges such as privacy and security.

On the long term, we see major technology players such as nanotechnology, 5G–10G, and others as disruptive technology that calls for distributed collaborative intelligence making sensing more intelligent, moving services from edge of the network to be distributed between End sensing node and cloud intelligence, inter-collaborative cloud will remove global barriers and finally solutions and services will also be hosted by a distributed providers (as a result of consolidation among service providers, carriers, etc.).

1

Introductions and Motivation

1.1 Introduction

This book will provide the reader with a quick overview of the Internet of Things (IoT) as the next technical revolution of the Internet and address its impact on digital life and business process empowered by emerging Cloud Services. The IOE – Internet of Everything is a DOT (Disruption of Things) and agent of change transforming from Internet of People to Internet of M2M (machine-to-machine) to IoT and IoE. Key benefits include improved business process efficiency, productivity, and quality of digital life. The market demand is moving from reactive on-demand remote monitoring and control (pull model) to more proactive services where the services will identify who you are and deliver what you want and when you want. Services will take on a new form empowered by insights driven from data collected in a given target applications.

This book is targeted to:

- Decision Makers, CEOs, CIOs, CTOs, and Senior Executives
- Communications and Networking System Architects
- Product and Marketing Managers
- Software and Hardware Engineers
- Consultants, Entrepreneurs, and Startups
- Service Providers of content, utility, security, entertainment, and other services
- Professors and Students
- Hobbyists.

1.2 The Book

1.2.1 Objectives

The objective of this book is to provide a simplified approach to comprehend the IoT innovative business model, value, and the impact it will have over our lives and businesses.

Collaborative Internet of Things (C-IoT): For Future Smart Connected Life and Business, First Edition.
Fawzi Behmann and Kwok Wu.

Hence, the objective will be analyzed from two aspects:

1. Improving quality of our lives
2. Improving business efficiencies.

Both (1) and (2) will contribute to having better insights that will lead to living in a smart environment and being part of a smart community/smart city.

This book will provide both a simplified and innovative business model for IoT and IoT domains and key applications. In addition, window of opportunities for IoT applications is rapidly growing and this book will address the need for a platform where diverse applications and solution can be developed. Finally, there is a need for system solution to collaborate for better insights and decision-making.

1.2.2 Benefits

It is the hope that this book will be a great reference and a tool for the reader and provide tangible benefits that include:

- Prepare the readers to embrace the revolution of Internet Technology with IOT that impact our lives, our business, and the environment we live in.
- Understand the evolution of IoT opportunities in terms of key applications that impact our lives, business, or the environment. Examples include Health & Fitness, Smart Building & Home, Video Surveillance, Track & Monitor, Smart Factor/Manufacturing, Smart Energy, and others.
- Comprehend the simplified approach to IoT Business Model consisted of Sensing, Gateway, and Services. Understand landscape of technologies and standards that enable creation of innovative time-to-market solutions and systems.
- Describe technologies and protocols that directly relate to IoT for IoT Architecture model (Physical, Virtual, and Cloud). Examples:
  - *Physical.* Sensors, radio frequency identification (RFID), micro-electro-mechanical systems (MEMS), wireless sensor networks (WSN), global positioning system (GPS), ZigBee, near field communication (NFC), Wi-Fi, and 3G/4G
  - *Virtual.* Software Defined anything (SDx), network functions virtualization (NFV), IPv4/IPv6, geographic information system (GIS), and body area network (BAN)/local area network (LAN)/wide area network (WAN) (body area network/local area network/wide area network)
  - *Cloud.* Big Data/Analytics.
- Acquire deeper the understanding with series of requirements of key applications and how they are implemented.
- Leverage a smart IoT platform approach to facilitate the development of applications and enable communications among systems and solutions in a collaborative

manner (C-IoT). Smart IoT as characterized by the 8A's stand for Automated Remote Provisioning and Management, Augmented Reality, Awareness of Context and Location, Analyze, Take Action, Automate, Anticipate, Predict, Autonomous, and Attractive. The 7S's stand for Simplicity, Security, Safety, Smart, Scalable, Sustainable, and Sleek Appeal.
- Acquire deeper understanding of Secured IoT products and systems. Provide examples and tips for do-it-yourself (DIY) IoT solution.
- Make the reader aware of future challenges facing IoT.
- Inspire the reader to think and innovate on unlimited IoT opportunities by sharing examples of future trends.

1.2.3 Organization

This chapter introduces a simple innovative model for Collaborative Internet of Things (C-IoT) and a new way of looking at the market via domains and applications. The book introduces simplified segmentation for C-IoT consisting of three domains: Individual, Industrial, and Infrastructure. Applications/Solutions/Systems can impact Individual, Industry, or Infrastructure or all. The C-IoT model in its simplest form consists of sensing, gateway, and services. Gaining insights from application/solution/system will result in driving better results in each domain. Spanning applications/solutions/systems to more than one domain create a higher value for a better strategic decision.

The core of this book will highlight a series of cases spanning from the requirements to the solution for present market and business opportunities and exploring future opportunities.

Chapter 2 describes the landscape of C-IoT and highlights relevant technologies and standards of the C-IoT model. In addition, it illustrates the application requirements by selecting some of the C-IoT applications spanning one or more of the three domains: Individual, Industry, and Infrastructure.

Chapter 3 describes the implementation of some C-IoT applications progressively from personal consumer to home to industrial (business) and smart city (infrastructure and communities) levels to deliver sustainable smart living and smart environment that help optimize business process efficiency and improve quality of life.

Chapter 4 describes the need for a unified smart C-IoT software platform that enables one to build and deploy smart C-IoT product, systems solutions, and services for different vertical markets in a quick time-to-market fashion. This will be described including the 4A's and 4S's that characterize smart C-IoT products. The 4A's stand for Awareness of Context and Location, Analyze, and Take Action, Automate and Autonomous, and Anticipate. The 4S's stand for Simplicity, Security, Smart, and Scalable. The chapter ends with covering the Secured IoT.

Chapter 5 provides examples of do it yourself (DIY) kits aimed on bringing the concept, approach to hands-on experience that will inspire innovative thinking in exploring untapped opportunities that will improve the quality of humanity, and business efficiency in general.

Chapter 6 addresses the emerging new wave of new devices and technologies such as wearable/mobile and cloud technology (local, public, and inter-cloud), analytics, and social media as key building blocks of C-IoT distributed intelligence. In addition, the chapter examines the C-IoT impact on our digital lives and businesses and some of the future challenges such as privacy and security.

Chapter 7 presents the readers with final remarks and tips to inspire them to innovate and deliver differentiated solution that contributes toward improving quality of lives and improves business processes.

1.2.4 Book Cover

1.2.4.1 Title of the Book

The title is carefully constructed that is depicted by the image on the front cover of this book as well as the brief text in the back cover.

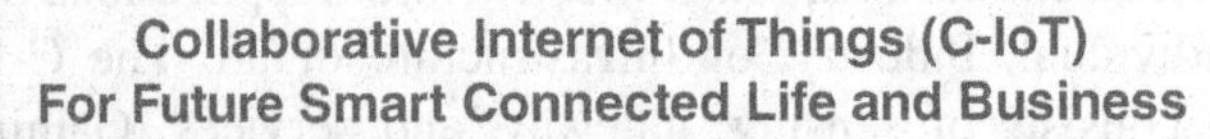

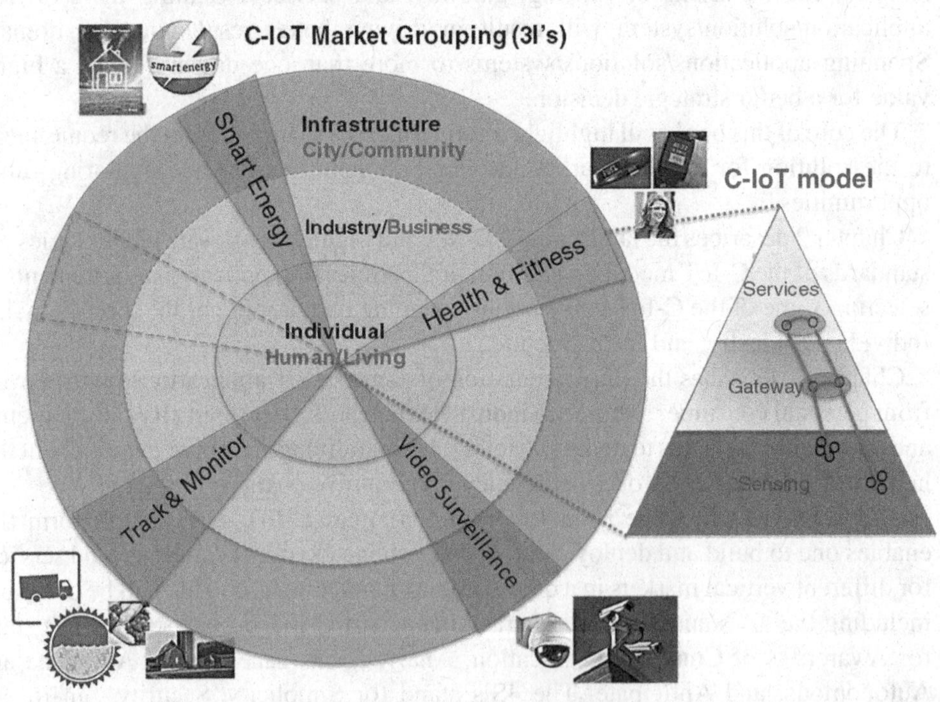

The title reflects the direction of the book focusing on the *Future* and addressing the need for a C-IoT that represents intelligently interacting and collaborating with current IoT solutions and systems to gain improvement and greater benefits that will have a positive impact on the *quality of our lives and our business in terms of process efficiency.*

The title can be depicted by two diagrams, which consists of:

- C-IoT domains and applications represented by the three circles (Individual, Industry, and Infrastructure) and examples of apps represented by triangles that cross the circles (examples of apps: Health & Fitness, Video Surveillance, Track & Monitor, and Smart Energy).
- IoT Model represented by the triangle with three functions (Sensing, Gateway, and Services).

The C-IoT in the title reflects the impact C-IoT has in breaking silos exist today in vertical markets by enabling communications and collaborations of solutions and systems within each market and across IoT domains: Individual (Human), Industry (Business), and Infrastructure (Government/smart city).

C-IoT impacts Individual, Industry, and Infrastructure and contributes in improving quality of life and/or improves business operational efficiencies.

- On the Individual C-IoT, a consumer is interested in buying a Nest thermostat because it saves energy without the hassle of constantly tweaking the thermostat by hand. A consumer is interested in fitness trackers such as Fitbit or Jawbone UP because visualizing the progress reinforces the behaviors that will make him or her fitter and view and act on video surveillance that is triggered by detecting a presence of a potential intruder.
- On the Industrial C-IoT, Internet solution providers (spearheaded by GE), focus on making industrial processes more efficient. For example, Industrial sensors for machines can be used to reduce the downtime, saving money. Business owner (referred here as CIO) can view information on the fly from anywhere using, for example, mobile device and conduct other queries and scenarios and then make a decision.
- On the Infrastructure IoT side, we see many initiatives on selling technology to city administrations that improves the operations of lighting, parking, water supply, garbage collection, and improve city data networks.

1.2.4.2 Walkthrough Example

C-IoT solution spans sensing, gateway, and services represented by the vertical line within the triangle (also referred to as a point solution). C-IoT is intelligently connecting two point solutions within the triangle. For example, connecting iWatch to not only manage health and fitness but also control remotely video surveillance at home or open/close garage door, or adjust temperature as person enter the home or car, and so on.

Imagine the following scenario: an elderly lady wearing an iWatch and she fell on the ground and broke her legs at home. She also had food being cooked on the stove but cannot reach it and shut it off. With C-IoT solution, automatic scan of video images and other sensors around her will generate multiple alerts that will be transmitted in real time to EMS (emergency medical services). EMS personnel will diagnose remotely the situation and based on assessment, other alert may be sent to fire department and police to protect the house from potential risk being caught on fire. With a severe condition of broken legs, EMS will send alert to a hospital closer to the home of the elderly. EMS, Police, and Fire vehicle will be dispatched finding the shortest path and interacting with smart street poles to reach the home of the elderly. With smart poles, First Responders will have access to change the traffic light, send messages to all vehicles on along the path to clear the road. And real time alert will be sent to the utility company to turn off the stove remotely.

We can see clearly from this example of a collaborative – IoT solution impacting three domains: Individual (elderly lady), Industry (Utility, surveillance alarms, hospital/medical services, and insurance), and Infrastructure (EMS, Police, and Fire). The end result is efficiency, faster services, which leads to an improved quality of livings, safety, and operational efficiency.

1.2.5 Impact of C-IoT

IoT is bringing a new level of coverage and information that were not possible before due to technological advancement and scalability DOT. IoT via variety of sensors will increase smartness and intelligence of information gathering that will impact a better decision and results.

We have embarked into the next era of technology evolution, after PC-era and post mobile-era. With the rapid evolution of mobile applications and generation of big data, there will be need for series of analysis to make sense of the data and present the user with factual and filtered information so that sound informed decision can be taken and communicated. This level of communicating quality results will open the window to a new level of smart communication that will lead to a collaborative networking supported by social media and break down barriers for global and effective and intelligent global communications.

IoT will enable convergence of several technologies and consequently will impact the overall architecture of the network so that we envision a common architecture platform for most of the vertical markets and adaptation and customization through applications and special devices to address the specific needs of a given market. This in turn will accelerate time-to-market of applications and services in many markets.

The impact is huge in terms of explosive growth of Internet traffic, increase of broadband speed, and the emergence of multimedia class of applications. This has led to the migration from client/server to cloud computing and services.

The future calls for a better insights not only through Compute engine for data processing, but also with analyzing workflows, advanced analytics, stream processing, and business intelligence that is GPS-based and visual analytics that is GIS-based. This helps in improving judgment and speed process of taking a better decision.

Security has become and remains to be a key challenge to secure access of content at anytime and anywhere.

We can see users equipped with remote access and control capabilities to manage the home environment (energy, safety, and security). This will contribute toward improve efficiency, reduce stress, and improve quality of life. In the area of health and wellness, consumer will have access to gadgets and devices that help track indicators about their physical condition and well-being so that a proper proactive course of action can be taken.

For example, when buy a car you have the option to activate smart services after purchase. Not sure if you want the model with the turbo? No problem, we offer a telematics service where additional muscle can be delivered to your vehicle when you need it and turned off when you do not. That is IoT meets cloud and mobile – the convergence of disruptive technologies doing what they do best: rewriting the traditional business rulebook.

Of course, it is easy to scoff at these examples, but are they really so crazy? Today there are real-world cases where savvy businesses are disrupting markets with practical applications in our connected world.

Take, for example, a UK company called Insurethebox. This business has developed a telematics device that can be fitted to a car to monitor driving behaviors. Insurethebox has wrapped the hardware with software and services that allow customers to purchase insurance according to the number of miles they expect to drive in a year and then monitor usage from a personal portal. Furthermore, if they drive safely they will be rewarded with bonus miles that help keep their premiums down. That is the IoT meets cloud meets gamification – again, highly disruptive.

What is exciting about examples such as Insurethebox is not so much the technology, but the opportunities created. The telematics hardware is only a means to an end, the real deal is the additional monitoring apps, APIs (application programming interfaces), and services, which when taken all together allow the company to deliver new and compelling value.

Collectively, the Smart Connected Digital Life (Smart Home, Office, Factories, Hospitalities, Transports, etc.) will contribute to a better quality of life, greater business efficiency, and new venues to generate revenue. This book will touch upon many of the services that can now be integrated in the Smart Connected Digital Life.

C-IoT with collection and processing of data can provide insights, which will lead to an improvement:

- For an individual living in a smart home, when he opens the door, an action will automatically disarm the alarm.
- For a transport industry, when driving a smart car, a collision is detected by a Bluetooth dongle reading vehicle data and automatically calls the emergency services.
- For a smart city, with smart parking meters, the sensors embedded in road surface can help find an empty spot and enable you to park quickly without frustration.

The combination of cyber-physical and social data can help us to understand events and changes in our surrounding environments better, monitor and control buildings, homes and city infrastructures, provide better healthcare and elderly care services among many other applications. To make efficient use of the physical-cyber-social data, integration and processing of data from various heterogeneous sources is necessary. Providing interoperable information representation and extracting actionable knowledge from deluge of human and machine sensory data are the key issues. We refer to the new computing capabilities needed to exploit all these types of data to enable advanced applications as physical-cyber-social computing.

1.2.6 Summary

The Internet has broken many physical barriers connecting people, companies, and communities of interests worldwide representing different markets.

Sensors generate data, data produces knowledge, knowledge drives action. Thus, making sense of the data that those devices generate creates the real value for users.

The introduction of smartphones has generated a wide acceptance and adoption by business, consumer and general population. Today, we see high growth in mobile traffic than landline. Increase longer life battery, introduction of tablets and growth of ecosystem for mobile applications will cause a gradual decline in the growth of PCs and rapid end of life of desktop equipment.

Miniaturization, progress with energy issues and cost reductions have resulted in rapid growth in deployment of networked devices and sensing, tightly connecting the physical world with the cyber-world as well as interconnected humans bringing along them virtual social interactions.

The scalability of Internet, advancement of wireless technology, accelerated growth of mobility, introduction of wearable devices, lower cost of sensing technology, lower cost of embedded computing, advancement of storage technology and cloud computing and services and the emergence of the area of analytics provide a perfect storm to the IoT across many markets. With C-IoT, more data are captured, tracked, analyzed in relation to relevant data from other applications that will be reviewed for taking a better and informed decision.

The number of devices connected to the Internet already exceeds the number of people on earth and is estimated ranges from 50 to 200 billion devices by 2020. The resulting system called IoT incorporates a number of technologies including WSN, pervasive computing, ambient intelligence, distributed systems, and context-aware computing. With growing adoption of smartphones and social media, citizens or human-in-the-loop sensing and resulting user-generated data and data generated by user-carried devices have also become key sources of data and information about the physical world and corresponding events. Data from all these sources will result in tremendous volume, large variety, and rapid changes (velocity).

The big data is not as in the past consists of numerical data and taking into account the internal of an organization but rather consists of a host of data from diverse sources and of all types – audio, video clips, social media and other forms spanning the internal, the environment and social culture and of course the supply chain.

Such information was not feasible before but thanks to technological advancement and IoT to sense more data that were passive objects. As a matter of fact, Big Data is becoming a potential gold mine and a great business to the point that contrasting to Big Data, a new category called small massive data is emerging providing a focused prospective to make decision and drive innovation and strive for a better quality of life. Thus, the CIO may be traditionally applied to Enterprise, but in the context of this book, with the introduction of IoT model, IoT platform to drive IoT innovative services, CIO is applied to all vertical markets small or large and furthermore, applied to IoT Services addressing the total supply chain for vertical market and crossing multiple vertical market in offering new class of IoT services that were not envisaged before.

With C-IoT, billions of devices and sensors, all communicating through the cloud and all feeding into a massive analytics solution to provide a complete picture of the individual or business, processes, and customers will be coming together.

Extracting value from your data requires integration of disparate tools. The challenge is to dramatically simplify big data processing and free users to focus on turning data into value with a hosted platform and an initial set of applications built around common workflows.

The C-IoT has the capability to collect and analyze feedback, report it to decision makers, enable them to take the appropriate action, and provide an analysis as to how well the actions taken worked.

The C-IoT can enable marketers to measure actual behavior in real-time rather than process-biased answers to unwanted or inadequate survey questions.

The C-IoT calls for delivering a Dashboard that allows CIO and user to visualize and get insights of big data in a graphical GIS-based form and relevantly located based on GPS. Interaction can be via mobile portable or wearable device phasing away from other traditional type of two-way communications.

This book describes IoT business consisting of three layers: sensing, gateways, and services. The definition is expanding from Internet of people, to IoT and now IoE. We are tapping into millions of objects that matter and bring them to Internet for tracking and extracting value that can bring a new dimension of information that was not possible before.

This book will assist the readers by providing supplementary material including case studies and code examples empowering a new level of collaborative innovation and creativity.

1.3 C-IoT Terms of References

This book introduces simplified and innovative concepts. This section provides definition and description of these concepts.

1.3.1 Introduction

1.3.1.1 IoT Landscape

The concept of IoT is not new. There are many early-IoT examples in the field of industrial control, process control, and telecommunication.

Fawzi Behmann, the coauthor of this book, was exposed to the concept of IoT in the late 1980s and early 1990s when he led a team of architects and developers into the design and implementation of a first computerized telecom alert system for a global carrier telecom facilities conforming to ITU TMN – Telecommunications Management Network Standards M.3000. The system that consists of network of computers provides three functions: Real time monitoring of equipment and systems, control based on fault analysis, provisioning, and performance, and supervisory of assessing impact on traffic and quality of services. Facilities alarms were monitored in real time via data acquisition systems with hundreds of digital and analog points identifying change of status or crossing defined limits. Such a change could have minor or major impact on traffic and services, which could have resulted in activating control options such as traffic rerouting. The system had built-in security including user access, functional access, and device and location information. Figure 1.1 illustrates an example of early-IoT deployment in the late 1980s and early 1990s with the emerging IoT in the 2010s.

Figure contrasts the early adoption of IoT with today's emerging IoT. For example, applications implemented in the 1980s used private lines and private networks where as today, Virtual Private Networks and wireless technologies are used. Back then, computer networks were more expensive with a limited performance, memory and

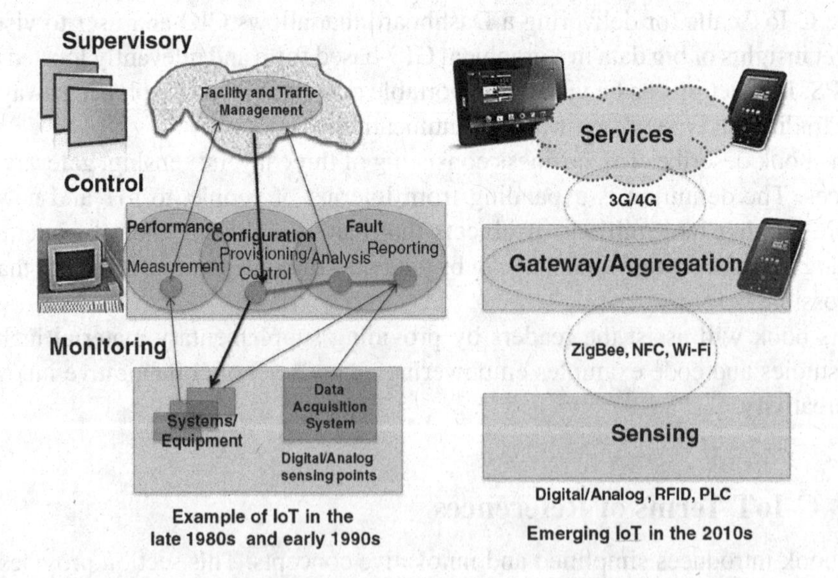

Figure 1.1 Contrast between early and current IoT Functional Model

storage. User interface was typically a CRT (cathode ray tube). Today, however, systems are more powerful, less expensive, and come with higher capacity memory, higher storage capacity and wireless connectivity supporting diverse user interface devices and digital monitors [1].

With the advent of Internet and advancement of enabling technologies for sensing, microcontroller, connectivity, and storage of big data, Solutions and services are taking on new meaning depending on where we are in the current IoT deployment and extracting value from the data collected. Although, sensing devices and control may provide an immediate and local value that can benefit individuals, small businesses and contribute to operational efficiency of enterprise, but sharing the data at a larger scale can benefit multiple vertical markets and can then generate a much bigger value to many large entities. Take, for example, wearable devices for fitness and health monitoring, they are of immediate benefits to individuals, but if a person pay visit to a physician that may end up in medication and being in the hospital, this would involve physicians, hospitals, insurance coverage, prescription, and a host of other agencies. If they are interconnected and patient data is respected for privacy, a consolidation and consistency of data will now emerge. The person has ultimate access to all the data being collected, trends can be generated (connecting the dots), a faster service can be offered since authorized agencies have access and share the data. Insurance agency can respond if the prescribed medication is covered, and less headache on the part of the patient to deal with each of these agencies as before and a better resolution.

We start witnessing an emergence of a new level of services, some being offered by a utility company, some by a transport car services, others by a carrier (MVNO).

It becomes a question that has control or influence on critical technology, infrastructure, and ability to collaborate with other entities to develop a new class of services.

1.3.1.2 IoT and Internet of Everything (IoE)

We are currently experiencing the IoT, where millions of new devices are regularly being connected to the Internet. As these "things" add capabilities such as context awareness, increased processing power, and energy independence, and as more people and new types of information are connected, we will quickly enter the IoE, where things that were silent will have a voice.

When smart things everywhere are connected together, we will be able to do more and be more. This is the IoE, a paradigm shift that marks a new era of opportunity for everyone, from consumers and businesses to cities and governments.

IoE is changing our world, but its effect on daily life will be most profound. We will move through our days and nights surrounded by connectivity that intelligently responds to what we need and want – what we call the Digital Sixth Sense. Dynamic and intuitive, this experience will feel like a natural extension of our own abilities. We will be able to discover, accomplish, and enjoy more.

Powerful smart phones are a natural tool to deliver IoE experiences. And wireless networks, a fundamental layer of IoE connectivity, are integrated around the globe.

Those elements are ready for everyone to leverage, and Qualcomm is making that easier than ever. Qualcomm visionary solutions are to deliver the connectivity and communication needed to support the IoE opportunity industry-wide.

Cisco defines IoE as bringing together people, process, data, and things to make networked connections more relevant and valuable than ever before – turning information into actions that create new capabilities, richer experiences, and unprecedented economic opportunity for businesses, individuals, and countries.

For simplicity, this book assumes that IoT will include both things and people.

1.3.2 *Need for IoT Framework*

Today, the market is flooded with information about IoT: Articles, papers, books, forums, and even ecosystem organization facilitating joint collaboration. This book will take a different approach by focusing on IoT connectivity that matters. In this context, the book will serve as good reference and guide to executives and leadership in business, colleges, and homes focusing on discernment on what matters in IoT. This book also intends to empower the reader with potential possibilities for a new innovative product, solutions and services and thus advancing IoT to a new level.

In order to achieve this, the book will focus on IoT for connectivity that matter and that result in better benefits such as improvement of efficiency of business and improving the quality of our lives. For example, IoT connectivity of things such as pens, erasers, staplers, hammers, and nails will not bring such value as that of IoT connectivity to control things such as doors, lights, and dishwashers. In this context, programming and managing a cluster of sensors for air condition for the home or enterprise could result in a lower operational cost. Smart parting where an individual is equipped with a smart parking map identify where is the available parking slot and can reach the available parking spot by visually looking at the green light on top of a parking spot seen from a distance. This can easily save time and avoid frustration.

This book will provide a definition and detailed description of the C-IoT model highlighting enabling technologies and standards; providing cases and examples in for several business applications highlighting requirements and implementation spanning from Infrastructure to Industry to Individual. C-IoT platform and other consideration will be highlighted. This book will also examine the C-IoT impact on our digital lives and businesses and some of the future challenges such as privacy and security.

In subsequent chapters, this book will cover:

- Technology and standards driving IoT
- Requirements for IoT in key areas
- Solutions examples in the areas of five key vertical market areas
- Impact on business and our lives.

1.3.3 C-IoT Domains and Business Applications Model

The core concept of the book is defined by two terms: C-IoT domains/business applications and C-IoT business model.

This book is taking the reader to a new level of IoT called C-IoT and thus providing differentiation and highlighting greater opportunities for creativity and innovation in the areas of personal quality of lives and advancement in business processes. This is accomplished by C-IoT domains/business applications providing a simple view of the market by identifying three principle domains represented by the three-part circles (Individual, Industry, and Infrastructure) and a sample of business applications represented by triangles that traverse these domains. The second concept is a simple business model C-IoT represented by the three-part triangle (Sensing, Gateway, and Services).

1.3.3.1 C-IoT Domains/Business Applications

The C-IoT defines three main domains

- Individual for smart living.
- Industry for business efficiency.
- Infrastructure for smart communities and cities.

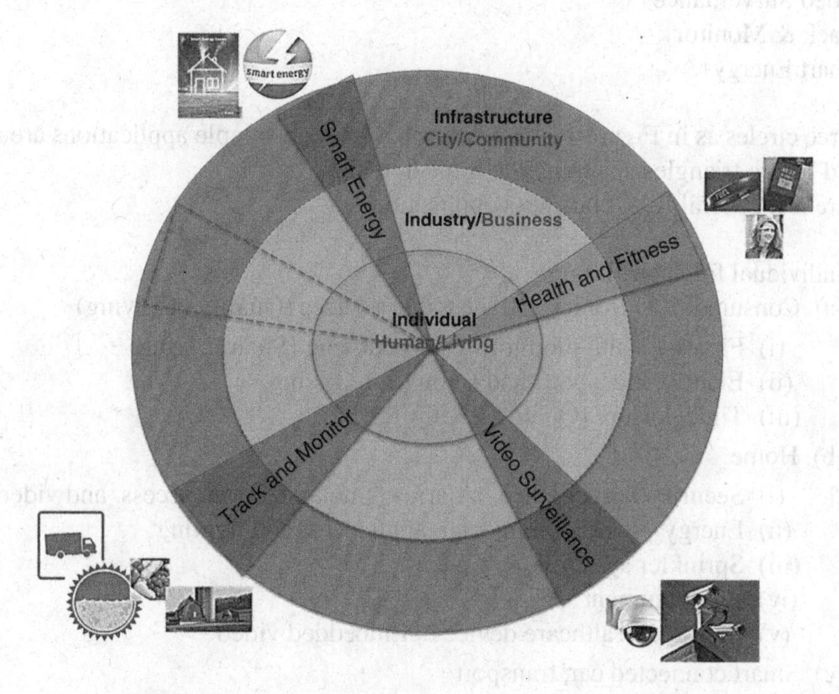

Figure 1.2 C-IoT domains and business applications

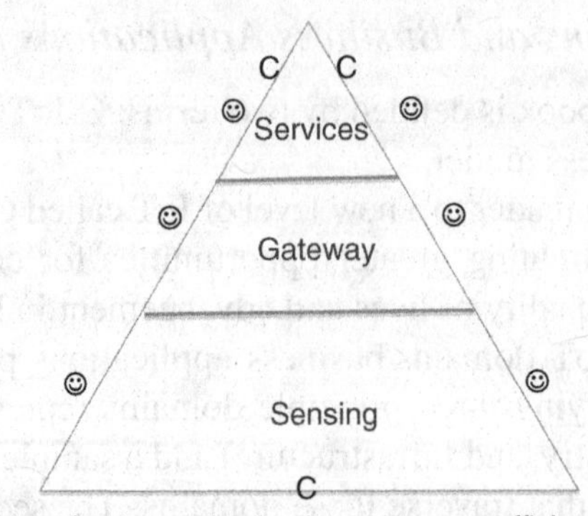

Figure 1.3 C-IoT model

Any business application may span one or more markets.

For example, in the same figure, the following business applications span the three domains:

- Health & Fitness
- Video Surveillance
- Track & Monitor
- Smart Energy.

Three circles as in Figure 1.2 represent domains and sample applications are represented by the triangles spanning the three domains.

Here is a general list of business applications:

1. Individual for smart living
   (a) Consumer Electronics and wearable devices (Quality of Living)
      (i) Fitness/health monitoring and tracking (Smart Living)
      (ii) Elderly, kids, pets, and so on (Safe Living)
      (iii) Time, leisure (Quality Living)
   (b) Home
      (i) Security/Surveillance – Alarms (smoke, motion, access, and video)
      (ii) Energy – meter, thermostat, appliances, and lighting
      (iii) Sprinkler system
      (iv) Entertainment
      (v) In-home healthcare devices – Embedded video
   (c) Smart connected car, transport
      (i) Engine diagnostics
      (ii) Traffic info/mapping

        (iii) GPS/GIS Location-based services
        (iv) Autonomous driving
2. Industry for business efficiency
    (a) Several markets are associated with Industry.
        (i) *Factories*. Automation, machine vision, robot, and machine learning
        (ii) Smart Manufacturing
        (iii) Agriculture (Produce, Livestock, etc.)
        (iv) *Health*. Clinical labs, health monitor and diagnostics, treatment, and health insurance
        (v) *Construction*: Smart building, heating, ventilation, and air-conditioning (HVAC), lighting control, and energy
        (vi) Smart Retails (Shops and Hospitality), Smart ePoS, vending machine, signage, ATM/Kiosks/Vending machines, and so on
        (vii) Smart Energy Grid, water, waste, pipeline, refinery, and air
        (viii) Smart Environment – Surveillance, Air and water quality
        (ix) Communications
3. Infrastructure for smart communities and cities
    (a) *Public Transportation*. Automotive, smart vehicles (V2X), trains, planes, buses, trucking, and cargo, Aerospace/Mil
    (b) *Public Highways*. Tolls, lighting, smart parking, meters, and so on
    (c) *Public Safety*. Police, fire, ambulance, surveillance, and drones
    (d) Disaster Management
    (e) Smart education, Security, and Defense
    (f) *Smart City*. Sustainable living and smart environment.

1.3.3.2 C-IoT Business Model

C-IoT Business Model is represented by the triangle in Figure 1.3 and consists of three layers: sensing, gateway, and services.

The model shows connectivity at each of the functional areas represented by the letter (C). User interface can also be at any of the three layers using mobile/wearable device and is represented by the smiling face symbol.

Today, many of the IoT business applications and solutions are point solutions.

By contrast, pre-IoT model represents a disjointed piece of the triangle and connectivity and interaction would require efforts and tools.

Example 1 If, for example, an IoT device such as iWatch, which has many sensors, can process data for local view and connect to the cloud for trends and analysis. This would be represented as a point solution with many sensors to provide, for example, reading on health and fitness reading. Another example would be to operate to open/close a garage door. This is represented by a second IoT point solution as shown in Figure 1.4.

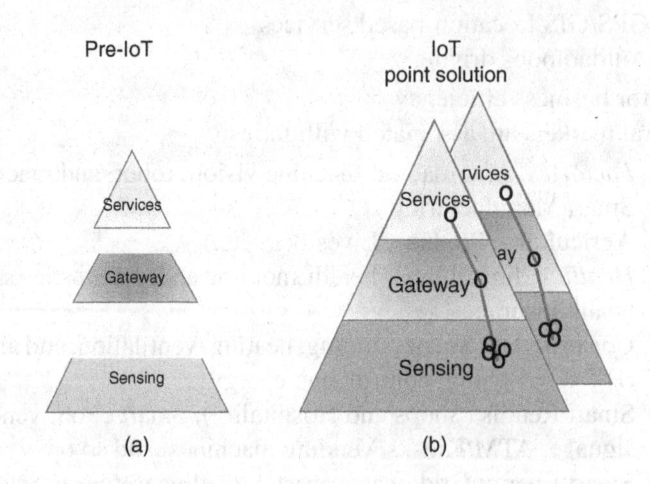

Figure 1.4 (a) Pre-IoT and (b) current IoT point solutions model

Figure 1.4 illustrates both pre-IoT and current IoT Point solution.

Example 2 If, for example, iWatch is to not only manage health and fitness but also control remotely video surveillance at home or open/close garage door (two other point solutions) then iWatch needs to intelligently interact, connect, and collaborate with these two other point solutions. This is represented by the horizontal connection between point solutions and is referred to as C-IoT. This is depicted by connecting

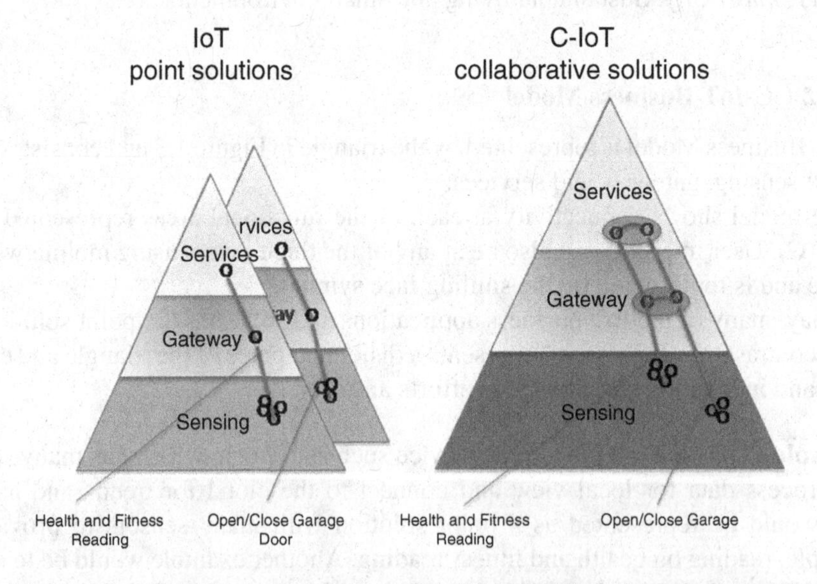

Figure 1.5 (a) IoT point solution and (b) C-IoT collaborative solution model

two point solutions in the triangle both at the gateway level and at the service level. Of course, the example can be more complex than that.

This C-IoT integrates intelligently and connecting two point solutions within the triangle as shown in Figure 1.5.

In the near future, there will be transformation toward C-IoT between two or more point solutions/systems providing greater insights that would result into better decision.

But such solution/services can impact any, some or the entire three domains: Individual, Industry, and Infrastructure. For simplicity, we highlighted three market cases that address our needs and wants: Health & Fitness, Video Surveillance, Track & Monitor and Smart Energy. More examples are provided in the book.

Example 3 Video Surveillance across Three Domains (Individual, Industry, Infrastructure) The following example shows a more progressive example of video surveillance business application crossing three domains: Individual (Apartment), Industry (Commercial), and Infrastructure (City Street).

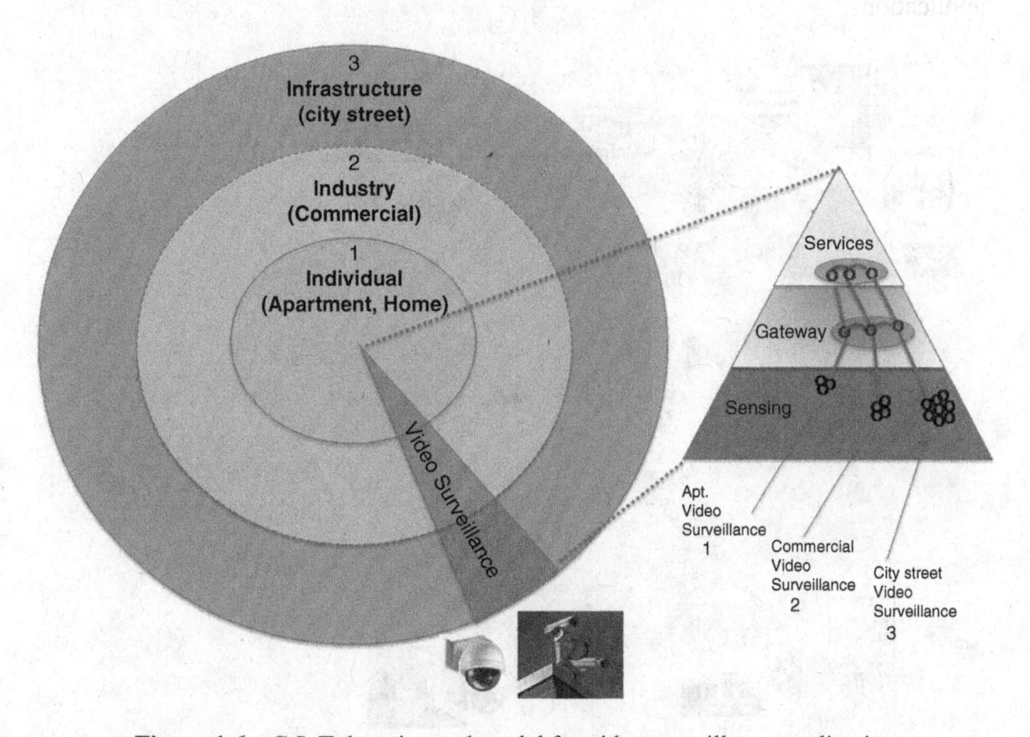

Figure 1.6 C-IoT domains and model for video surveillance application

A major fire has occurred in a multi-tenant building, which contains both apartments and commercial stores and restaurants. A fire alarm was sent from an apartment that has fire and was supported by video surveillance images detected by central alarm services. Shortly afterward, a fire alarm was issued from a restaurant caught on fire in the building and it was supported by a video surveillance images sent to the central alarm services. The central alarm services have sent an alert to the Fire department. However, because the building is in central downtown, the street video surveillance camera has also detected the fire prior from receiving the alert message from the central alarm monitoring services. Thus, in real-time all the first responder, commercial building and apartment issued alarm alert for action. Immediately, the emergency dispatching center was able to retrieve more information about the building and exact location of fire displayed on a GIS template and right fire truck and resources were dispatched immediately to the site to put off the fire in both locations.

Thus, with a C-IoT, and the support of the video surveillance at multiple locations representing individual (Apartment), Industry (Commercial Building), and Infrastructure (City Street), detailed information was correlated in real time, and proper action was taken to minimize damage and save lives.

The illustration of this C-IoT example is shown in Figure 1.6.

Example 4 Figure 1.6 shows a more general example of C-IoT for multiple business applications:

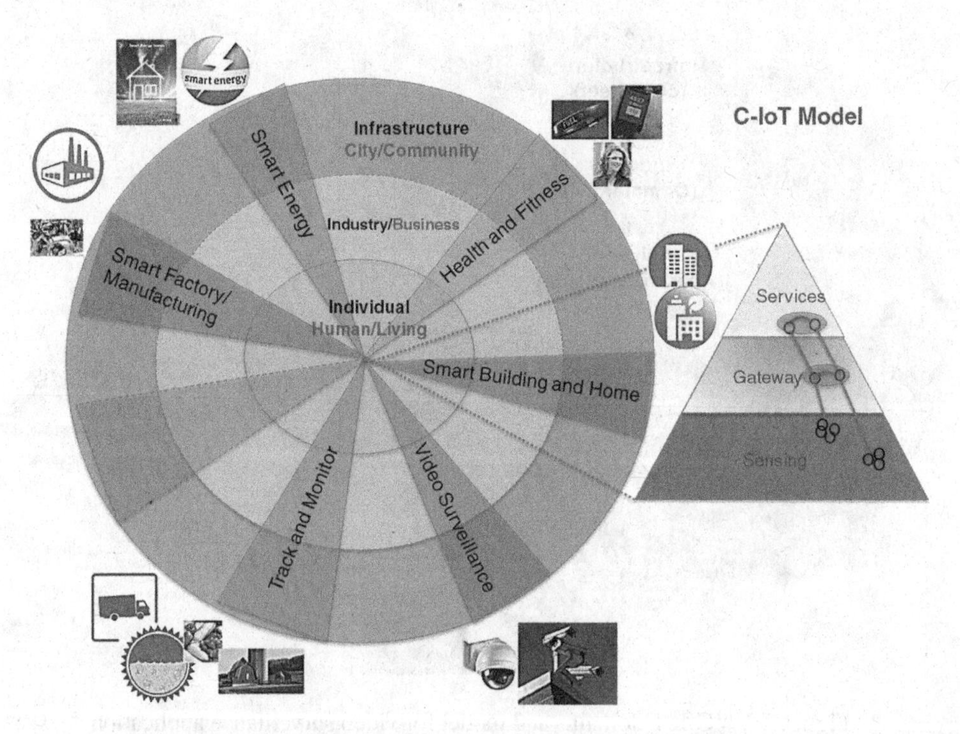

Figure 1.7 C-IoT general domains and sample applications

- Health & Fitness
- Smart Building & Home

- Video Surveillance
- Track & Monitor
- Smart Factory/Manufacturing
- Smart Energy
- Others.

These applications all cross the three domains: Infrastructure, Industry, and Individual. Figure 1.7 illustrates C-IoT domains and sample applications.

1.3.3.3 C-IoT and Internet of Service (IoS)

The C-IoT Business Model can also represent an integrated point solution represented by the connected lines spanning sensing, gateway, and services within the triangle. Multiple sensors feeding into Microcontroller in the sensing layer, which can be viewed and/or passed to the cloud layer via gateway, represent sensing. Data, trends, analysis can be carried out at the cloud layer and be viewed by a smart mobile device.

Motivation for C-IoT is the delivery of IoS (Internet of Service). Embedded things are connected to the cloud via smart IoT gateway software platform for delivery of IoS business. This will be covered in more details in Chapter 3. Thus, the IoS is a vision of the Internet of the Future (C-IoT) where everything that is needed to use software applications is available as a service on the Internet, such as the software itself, the tools to develop the software, the platform (servers, storage, and communication) to run the software.

1.3.3.4 Cyber-Collaborative IoT (C^2-IoT)

While IT experts have long predicted security risks associated with the rapidly proliferating IoT, this is the first time the industry has reported actual proof of such a cyber attack involving common appliances – but it likely will not be the last example of an IoT attack. IoT includes every device that is connected to the Internet – from home automation products including smart thermostats, security cameras, refrigerators, microwaves, home entertainment devices such as TVs, gaming consoles to smart retail shelves that know when they need replenishing and industrial machinery – and the number of IoT devices is growing enormously.

The "IoT" holds great promise for enabling control of all of the gadgets that we use on a daily basis. It also holds great promise for cybercriminals who can use our homes' routers, televisions, refrigerators, and other Internet-connected devices to launch large and distributed attacks. Internet-enabled devices represent an enormous threat because they are easy to penetrate, consumers have little incentive to make them more secure, the rapidly growing number of devices can send malicious content almost undetected, few vendors are taking steps to protect against this threat, and the existing security model simply will not work to solve the problem.

IoT devices are typically not protected by the anti-spam and anti-virus infrastructures available to organizations and individual consumers, nor are they routinely

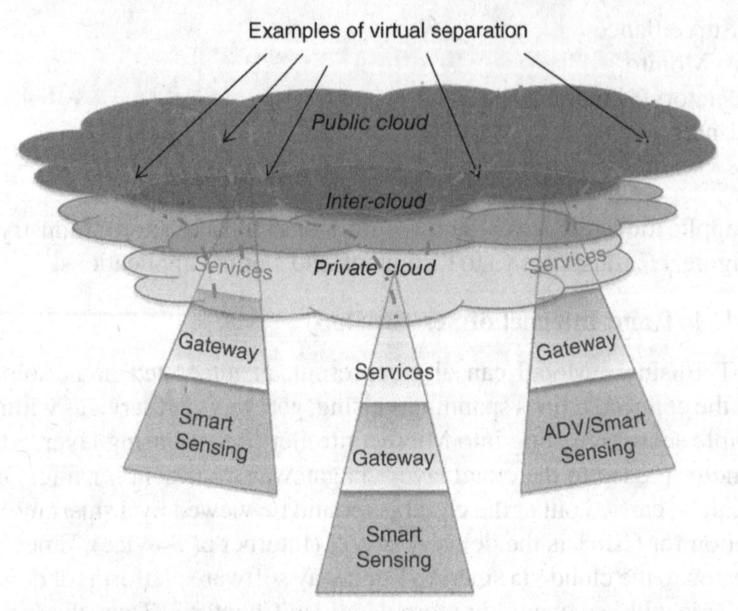

Figure 1.8 Cyber-collaborative IoT model

monitored by dedicated IT teams or alerting software to receive patches to address new security issues as they arise. The result is that Enterprises cannot expect IoT-based attacks to be resolved at the source; instead, preparations must be made for the inevitable increase in highly distributed attacks, phish in employee inboxes, and clicks on malicious links.

The view of the authors is that a long-term strategy would be required for the C-IoT. As the number of devices grows, so does the wireless chatter and noise. The chatter is building over all wireless communication channels: Wi-Fi, cellular, Bluetooth, NFCs, and others. There is no end in sight to the expected growth of wireless connected devices. Networks will need to be fortified and new methods of managing wireless traffic are being considered. Enriching wireless network traffic with rich context and history allows for dynamic network traffic prioritization based on the profiles of your customers. Always make sure that your most important customers always get the best Quality of Service, and know immediately when quality degrades.

The strategy to address Cyber-C-IoT is shown in Figure 1.8.

Security measure would be required both in the gateway and cloud areas.

1.3.4 Roadmap of IoT

1.3.4.1 Evolution Roadmap of C-IoT

Figure 1.9 represents the vision of C-IoT roadmap transitioning from IoT point solution in 2010s to C-IoT in the 2020s to Cyber-C-IoT in the 2030s and beyond.

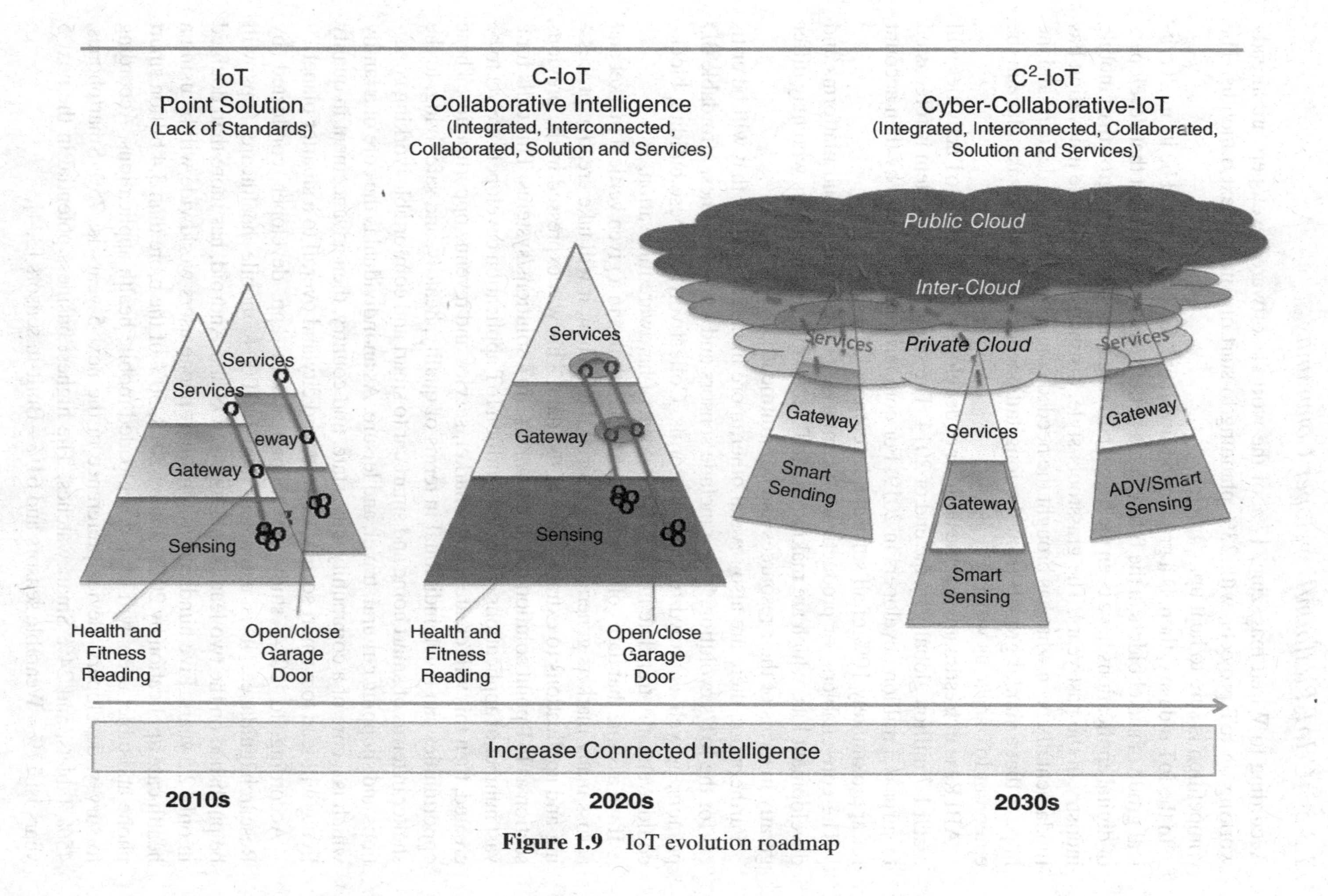

Figure 1.9 IoT evolution roadmap

1.3.5 C-IoT Platform/Developer Community

According to Wired magazine, 17% of the world's software developers are already working on IoT projects, with 23% planning to start one in the next 6 months. The competition is sure to heat up.

In the IoT industry, there is a growing notion that "developers" will be its most critical group of stakeholders in the future. This belief stems from the fact that developers of digital applications have been the biggest driving force behind the growth of mobile industry in the past years. The grassroots-style, distinctly speculative innovation has fundamentally been what has brought the needed content into mobile devices, and the hope is that a similar growth-from-the-long-tail phenomenon would take place in the emergent IoT space as well.

ABI Research estimates that the number of developers involved in IoT activities will reach 1.7 million globally by the end of 2014. The size of the ecosystem is forecasted to surpass 3 million developers in 2019. For comparison, this would by then account for approximately 10% of all software developers [2].

The core enablers for productization comprise purpose-built cloud platforms and development kits, which are making the IoT accessible to developers who may differ greatly in terms of their resources and commitment.

Furthermore, there are also several other, more indirect enablers that will be critical for the IoT's evolution. These include sensors and sensor engines, affordable 3D printers, as well as crowd funding platforms. Collectively, all these building blocks could eventually translate into a perfect storm of hardware innovation.

If we assume that data collected off connected things in a given vertical market and across many markets generate a huge set of data. Then, it will take creative processing and applications to extract value of the data. This was expressed in the previous section as IoT point solution moving toward C-IoT solutions/systems. This calls for a vast number of applications. Thus, 3–5 million IoT application developers are yet to be created, then this will create a huge market, service and revenue opportunities. These opportunities can touch individual in terms of quality, healthy, and safe living, to the shelter in terms of smart home and smart car to smart and comfortable working conditions and shopping cultural, travel, and leisure. As an individual is the core of a family which is a core of a community, city, state, and country, then advancement in quality living will lead to a smart society, culture, and city and overall to a smarter planet.

According to 2014 study on mobile health app developer conducted by Research2guidance, it is estimated that 100 000+ mobile healthcare apps will be published on the two leading platforms, iOS and Android, has more than doubled in only 2.5 years. Five hundred million smart phone users worldwide will be using a health care application by 2015, and by 2018, 50% of the more than 3.4 billion smart phone and tablet users will have downloaded mobile health applications. According to survey, developers' device preference in the next 5 years is: 75% Smartphones, 45% Tablets, and 24% Smart watches. The highest business potential in the next 5 years is: 77% – Wearable sensors and 61% – Built-in sensors [3, 4].

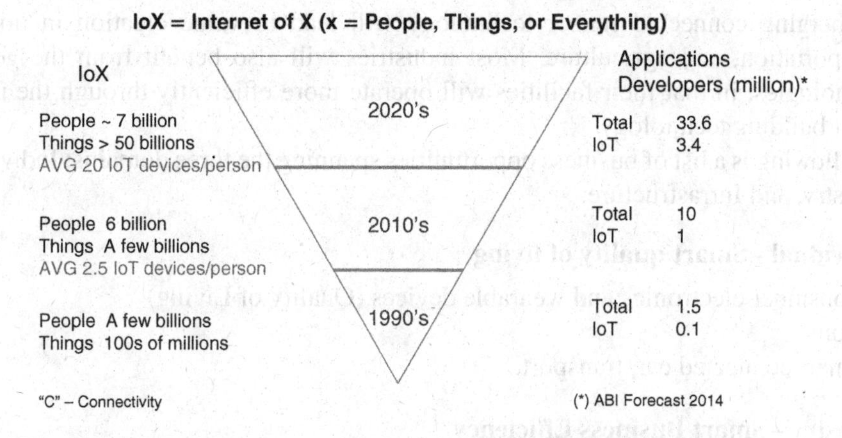

Figure 1.10 IoT contrasting IoE with IoT developers and apps

Figure 1.10 shows a composite picture of IoE (People and Things) contrasting with IoT developers at different time periods 1990s, 2010s, and 2020s.

1.3.6 C-IoT Opportunities for Applications, Solutions, and Systems

The Internet of People runs to billions of devices already. The IoT will involve ubiquitous smart objects that sense and communicate directly over the Internet creating more and better data. C-IoT is ubiquitous IoT or Collaborative IoE engaging things and people. C-IoT solutions/systems consists of smart sensing network with IP address assigned for smart objects with estimated tens of billions of smart objects, microcontrollers, potentially billions a year with sensor interfaces and wireless interfaces and connectivity to cloud for data analysis, storage, trends, and services.

According to Gartner, by 2020, component costs will have come down to the point that connectivity will become a standard feature, even for processors costing less than $1. This opens up the possibility of connecting just about anything, from the very simple to the very complex, to offer remote control, monitoring, and sensing, The fact is, that today, many categories of connected things in 2020 do not yet exist [5].

Business opportunities, which drive through the scale and usage of IoT technology span across Infrastructure, Industry, and Individual domains and include smart factories/manufacturing, smart grid, smart buildings, smart health, and fitness.

Emerging areas will witness rapid growth of connected things. This will lead to improved safety, security, and loss prevention in the insurance industry. IoT will also facilitate new business models, such as usage-based insurance calculated based on real-time driving data. The banking and securities industry will continue to innovate around mobile and micropayment technology using convenient point-of-sale (POS) terminals and will invest in improved physical security systems. IoT will also support a large range of health and fitness devices and services, combined with medical advances, leading to significant benefit to the healthcare sector.

Emerging connected sensor technology will lead to value creation in utilities, transportation, and agriculture. Most industries will also benefit from the generic technologies, in that their facilities will operate more efficiently through the use of smart building technology.

Following is a list of business opportunities spanning the three domains: Individual, Industry, and Infrastructure.

Individual – Smart quality of living

- Consumer electronics and wearable devices (Quality of Living)
- Home
- Smart connected car, transport.

Industry – Smart Business Efficiency

- Manufacturing
- Agriculture
- Health
- Building
- Retails.

Infrastructure – Smart communities/cities for sustainable environment and living

- Public transportation
- Public highways
- Public safety
- Disaster management
- Smart education
- Smart communities/cities.

To inspire innovators, entrepreneurs and business, following are set of examples.

1.3.6.1 Generic Examples of the IoT Benefits of Emerging Applications and Services

Future emerging applications and services will contribute toward time saving, operation costs, and streamline service provisioning.

- Connecting IoE to a network can quickly see the potential savings in time, cost, and labor.
- Transmitting information wirelessly at its source – from sensors, things such as vehicle, instrument, house, or business – and receiving valuable feedback not only automates manual processes, it helps to streamline service provisioning and billing.
- Multiply the number of connected devices, people and things you have, and you can see the powerful possibilities.

1.3.6.2 Examples of Business Applications and Solutions

Business Applications and Solutions are promised of improving communications among vertical markets. Benefits from device–device communications and remote monitoring/diagnostics will contribute toward cost saving and safety benefits.

- Applications and services connecting millions of diverse devices to a network, enabling two-way communication offered to key vertical markets such as heart monitors, vending machines, trucks, and turbine.
- *Oil and Gas Industry – Delivering Cost Savings and Safety Benefits.* Managing marginal well operations is labor intensive. Pumpers drive to each tank daily to check water and oil levels and equipment, including drilling motors. If levels are too high, pumps must be manually shut down to avoid spills and incurring fines and clean-up costs. Motor problems require a well to be shut down until repairs are made; both situations create a production loss. Pumpers can monitor tanks, motors, and other equipment anywhere from a PC or smart phone, 24/7.
- *Electronic Vehicle Charging Stations.* PEP stations needed a remote monitoring and reporting solution for its *electric vehicle charging stations*, which served building tenants. Cloud-based application provides a reliable wireless link to support real-time fault diagnostics. Benefits include: reduce call outs to fix faults, and receive reports on the utilization of all its equipment, which facilitated planning for new stations.
- *People Awareness and Recruitment.* Applications that provide a visibility for firms that are adopting device-to-device communication. Benefits include: improving prospects for direct sales, sales channel development, investment community awareness, and personnel recruitment.

1.3.6.3 Examples of Applications and Services for Supply Chain

Applications and Services for supply chain can provide critical information at different stages of supply chain providing cost saving and better services.

- *Integrated Remote Health Monitoring Services.* Traditional heart monitoring systems relied upon phone lines and even physical visits to deliver monitors and data; physicians waited for paper-based reports and technical analysis. An integrated remote monitoring system with a suitable wireless network provider deliver an effective solution that enables data to be automatically transmitted from heart monitoring devices to 24-h Monitoring Center, where credentialed technicians can interpret patient data, notify physicians of serious events, and upload reports to a web portal for physicians rapid access.

- *Improve Efficiency and Improve Service Time*. A system and application to remotely monitor medical systems. Benefits include minimize costly field repair visits and avoid downtime for vital patient treatment. Services enable timely problem identification and resolution. Benefits can be quantified in terms of recorded cost savings where recorded for both the medical systems and patients.

1.4 The Future

1.4.1 General Trends

We have witnessed a massive and ongoing revolution with computers and smartphones connected to the Internet. Smart mobile devices such as smartphones and tablet help individual to have access to information from one device. Through a set of apps, with such a device, an individual use it as a compass, navigate, take picture/videos, find gas stations, restaurants nearby, view emails, respond to telephone calls, play music, using it as a boarding pass, exchange contact information, pay a bill, and many other things all from one device.

At the end of this section in the summary section, we can see a potential use of future smart mobile/wearable device in the context of C-IoT.

While 2013 saw some remarkable advances in Micro-controller unit (MCU) and wireless technologies that further enable the IoT, 2014 has shown a forward move with the emergence of a more practical Internet of Useful Things. For some, this Internet of Useful Things promises a world where nearly every device continuously analyzes and transmits data about operational minutiae while filtering winnowing non-useful details.

The size of IoT opportunity by 2020 varies from one research firm to the other with an estimate over $10 trillion and over 50 billion connected devices.

The growth of connected things is accelerated by the massive growth of companion products to mobile devices, designed for consumers' increasingly on-the-go lifestyle.

The trend toward a more useful world of connected devices hinges on a new class of smart sensors. Sensor integration and sensor fusion will continue to be the watch words for these new sensor networks. In turn, more sophisticated sensor systems will rely on a new class of processors – blending low power and high performance data lower price point needed to enable widespread use. Expect more everyday devices to become intelligent and networked, billions of devices – thermometers, cars, light switches, appliances, homes, are being connected to the Internet – each with its own IP address.

In 2014, portable and wearable computing promises to introduce major shifts in how humans interact with computing devices and information, dramatically reduce the gap between immediate information and the person for whom that information is the most useful. Health and fitness buffs already wear monitors that record their heart rate and the distance they run, coupling that to a PC to analyze the results. Wearable wireless medical devices include accelerometers to warn of falls, and insulin pumps and glucose monitors for diabetics. Each of these devices can connect to a smartphone via Bluetooth, and can issue an alert to their physician by triggering a call over the cellular network.

While enhanced hardware solutions continue to emerge for smart sensors, systems designers are looking to surround users with a more natural environment that provides an immersive experience via interacting with IoT devices and the cloud. Ultimately, all these key technologies – the Internet of Useful Things, and wearable computing – serve the purpose of bending technology to meet real needs as efficiently as possible. With the continued evolution of these key areas, the industry is moving rapidly toward that objective. Companies are encouraged to enabling a standards-based IoT where billions of devices of all types and capabilities are connected through interoperable Internet Protocols and Web Services.

Let us look into the future and see where IoT is heading and how C-IoT will impact Individual, Industry, and Infrastructure (3I's) connecting to each other. The focus is to inspire the reader to invest now to profit from the future, explore some of potential opportunities that enable intelligent connectivity across the 3I's.

1.4.2 *Point Solutions*

There is increasing number of devices connected to the Internet in many areas of our lives. Here are some examples of point solution impacting Individual, Industry, and Infrastructure:

1.4.2.1 Individual

- Smart watches (e.g., iWatch) and wearable devices send, receive, and retrieve health monitoring information that is carried on the wrist or other parts of the body. Imagine if such device is connected to other smart devices such as smartphone, smart tablet, smart t-shirt, smart eye-glass, and drive more value!
- Crack-pot with sensor to raise or lower temperature to adjust to the timing as to when dinner need to be ready when the individual wants.
- Wearable glasses (e.g., Google Glass) with the ability to send and receive lots of different information in real time in the direction of where the wearer is looking. Imagine if the law-enforcement or military agent wearing this has added image recognition capability with cloud-based real-time criminal facial recognition, and has cloud-based voice recognition, natural multi-language language translation!
- Biometric to protect password – Fingerprint sensor enabled into your smartphone. Swiping your finger will act as a password.
- Context-aware Smart TV provides TV content and two-way communications capabilities via the Internet. Smart screens and Smart TV also have a camera that can detect the context (who is watching the TV). The concept of two-way interactivity is extended to mirrors, windows, and other screens in homes, classrooms, and businesses. They are already on the computers and mobile devices in the form of touch screen and tablet computers.

- Smart garage door opener and add smartphone to it to operate remotely. Imagine if the smart garage door recognizes the home-owners car arriving, auto-discovery and auto-authenticate for entry then auto-notify the smart-thermostat to set the AC to the desired temperature for the specific home owner!

1.4.2.2 Industry

- *Smart Parking Space that Waits for You*. Imagine the time spent and frustration in finding a free parking spot – all can be solved with smart parking!
- *Smart Ink*. Several manufacturers have developed forms of ink that enable electronic circuits to be printed on just about anything. This will enable consumers of print ads to interact with those ads – giving Marketing Information System feedback, transporting consumers to Web sites, and enabling them to request more information or even order products.
- *Smart Sports*. Smart Basketball, golf ball, tennis ball, smart rackets, smart bats, smart sneakers, and so on as a connected object that can be recognized once it is in a specific space. Imagine if the smart sport devices have capabilities to provide real-time data analytic and real-time feedback to suggest to the sport player how to improve!
- *Smart Clothing*. Sensors built into the fabric, or printed on, clothing will give health and performance feedback to medical personnel and athletes. Marketers will use such clothing to monitor physiologic responses to marketing content, product and brand variations, and pricing changes. Imagine if smart clothing is context-aware whereby you can even change color and transparency of the clothes based on own heartbeat and temperature as well as heartbeat and temperature of the other person near-by!

1.4.2.3 Infrastructure

- New context-aware LED bulb that is fully dimmable, and adjustable color (by context who, when, etc.) with simple API, which can be controlled remotely. LED headlamps are evolving beyond just a standard replacement for headlights in new vehicles. Increasingly, premium features such as side illumination and cornering headlights are appearing, thanks to greater reliability, smaller size, and lower cost of brighter LEDs. A similar example can be drawn for LED for buildings or for streetlights.
- *Smart Digital Signature*. Taking traditional forms of promotion into the future, these will be ads that communicate to potential buyers, enabling them to interact with the ads, request more information, and buy products directly from the ad. This can be applied for any industry as well can be considered as digital banners mounted on street poles for promotion, generate awareness or as an amber alert helping law and enforcement in the search of someone.

1.4.3 *Collaborative Internet of Things*

1.4.3.1 Individual IoT

Individual for smart living covers consumer electronics and wearable devices, smart homes, and smart connected cars that are connected to Internet and communicated with each other.

Health and Wellness

Wireless sensor technology is allowing us to easily monitor the critical life-signs of patients. Previously, the lack of suitable sensors, combined with an unworkable tangle of wires, has severely limited medical applications. Now patients can be monitored effectively for serious conditions such as sleep apnea, while in bed sleeping. For monitoring critical parameters such as heart rate, while a patient is undergoing strenuous physical exercise, the connection of wires is impractical. This has evolved where people can now experience IoT through wearable devices such as wrist watches that not only monitor time and manage texts and emails but also monitor the blood sugar in a person's system without the pain and suffering of a needle prick. The IoT has in fact conquered a lot of ground, even that of medicine and fitness.

BAN – Body Area Network

BAN is a network of devices applied on the person's body that monitors different parameters and provides feedback information to the person. In addition, the information can be relayed to other, or stored for trends and analysis.

At the bigger picture, tracking will be required for the cardiovascular system, respiration system, mental activities, and others to provide a wholesome view of an individual and help guard against common cold, to difficulty in having quality sleep.

But the trend is to help people to stay healthy by monitoring and tracking their health practice. Most of the band devices have capability not only to show the real-time results on a mobile device, but also send it to the cloud for storage. As we understand many of the symptoms of diseases and disorders can be traced in the body in advance before they occur. Such stored data can be used to obtain clues about such diseases or disorders before they occur. Also, in the future, the collection of such data about different individuals (with permission of the person) can be used to provide valuable insights to the scientists as to how the body reacts when faced with certain disorders or incurable diseases.

The US FCC has approved a 40 MHz spectrum allocation for low-power medical BANs.

Today, the wearable devices are in its infancy, but according to IHS research, the market for Wearable devices could be worth $6 billion by 2016.

Using Google Glass, a physician can now works in hand-free way, have immediate access to patient medical record from any place in the hospital. The device can be used in the operating room to help a surgeon monitor and control in real-time medical equipment without taking their eyes off the patient. It can also be used outside hospital at a site of an accident, where the paramedic can have access of medical information such as allergy about a person before they administer drug.

Robots can also play a role in the operating room in the future in assisting the surgeon on some of the difficult tasks and also can help in performing microsurgery when the surgeon is not in the room.

Home

Today, there are kits or gadget such as Raspberry Pi that enable you to connect your TV to the Internet for less than $25. Future TV will be Smart TV (Internet enabled and ready). For example, as you drive your car close to home, a signal can be sent out to open the garage door and adjust the thermostat setting in the home. As come close to the door, with video recognition will the door and various lights are turned on.

One example of smart device in the home is a "learning thermostat" by Nest. The device is using a series of sensors that monitors our daily schedule and adapt the climate as we come and go. The learning thermostat is promised to lower the cost of energy bill by 20%.

Another example by the same company called "Protect" is smart smoke detector that not only alerts the occupier via smartphone and has capability to call the emergency services. Advance model will have capability to communicate with home heating system and shut it down in the case of a fire or even a gas leak.

A more advance C-IoT is adding intelligence in communicating among various smart devices in the home. For example, when a person wakeup, a wearable device such as a wrest band can automatically send a signal to the smart thermostat to adjust the temperature, send a signal to the drapes to be opened, start the coffee machine, and so on.

Future homes will have modules of connected devices into a central system that can enable the consumer with smartphone or tablet to control all the functions in operating a home. With a touch of a button, user can exercise control over any system in the home such as security, climate control, or home entertainment. Turn on home projector, creating automatically the perfect ambience, sound system, and others can all be operated from a single app.

Smart Car

Having hundred's of sensors in a car make it more smarter, providing alerts when service is required, provide a higher degree of safety and gadgets that provide the driver with an ultimate enjoyable driving experience. Smart inter-car connectivity and communications help in avoiding collision. Connecting with the environment and infrastructure, a drive can be alerted of traffic blocks of potential bad weather conditions.

Smart Electric Car

Most electric cars have today had a management app enable us to communicate with the car, with a third party such as emergency services if there is a trouble or accident.

CAR – Heads-Up Display

Researchers are busy trying to solve problems, such as glare and windshield curvature, that affect daylight readability. Automotive displays need to deliver very bright images in small affordable packages, have a long lifetime and not generate heat. The aerospace has a continuous influence into the design – aerodynamics of cars and now in terms of instrumentation and guidance system. The latest is head-up display (HUD) technology, which was originally developed for fighter jets. The head-up projection system creates a very high brightness monochromatic image that appears only when the driver looks in a particular location. Such a display would allow the driver to view transparent images for daytime navigation information as well as night vision imagery by projecting the images off the windshield. Projecting information directly into the driver's line of sight allows people to process it up to 50% faster – due to shorter eye movement – and keep their attention focused on the road ahead. HUD technology is currently available on several high-end vehicles and is starting to show up in other segments. In fact, more than 35 vehicle models currently available in the United States have standard or optional HUDs. According to IHS Automotive, 9% of all new automobiles in 2020 will be equipped with HUD technology versus 2% in 2012. Sales this year alone is projected to climb 7% to 1.3 million units.

In the future, HUDs may alter the interior design of vehicles by eliminating the need for a center console equipped with a screen and a digital instrument cluster. A handful of Tier 1 supplier, such as Bosch, Continental, Delphi, Denso, Johnson Controls, Nippon Seiki, and Visteon, are currently exploring this issue. Engineers are scrambling to develop new HUD technology that can be cost-effectively mass-produced, while addressing issues such as noise, vibration, and harshness.

1.4.3.2 Industry

Industry for business efficiency covers several markets associated with industry such as factory automation, smart buildings, and smart retails.

With introduction of intelligence into equipment, IoT will revolutionize industries and making the business more efficient and provide customers with new services.

According to GE, today Industrial Internet (IoT for Industry) account for $32 trillion dollars. By 2025, it could reach to $82 trillion dollars of output or approximately one-half of the world economy. At GE Durathon battery factory in Schenectady, NY, the IoT helps the company collect data about processes going on 24 to 7. Through the 10 000 sensors on the assembly line and the sensors in every single battery, managers can instantly find out the status of production. They are able to share that information and data with coworkers in other departments.

According to O'Reilly's David Stephenson, technology at GE and at similar manufacturers has resulted in innovations never before realized. "The Internet of Things promises to eliminate massive information gaps about real-time conditions on the factory floor that have made it impossible to fully optimize production and eliminate waste in the past," he says [6].

Collaborative IoT and Manufacturing

Global competitive pressures are challenging industrial and manufacturing companies to drive inefficiencies out of their systems, manage workforce skills gaps, and uncover new business opportunities.

Factories and plants that are connected to the Internet are more efficient, productive and smarter than their non-connected counterparts. In a marketplace where companies increasingly need to do whatever they can to survive, those that do not take advantage of connectivity are lagging behind.

According to Rockwell Automation, only 10% of industrial operations are currently using the connected enterprise, which connects businesses to cyberspace to improve manufacturing functions. The machines were linked to FactoryTalk, software that lets the company's employees have remote access to both historical and real-time data and features production dashboards that provide a comprehensive picture of the whole system so they can monitor performance. All of this software and hardware resulted in faster time-to-market, improved asset utilization and optimization, lower total cost of ownership, workforce efficiency, enterprise risk management, and smarter expenditures. The new system reduced maintenance costs and downtime, as well as the capability to collect more data. In turn, this afforded them the ability to aggregate data and present dashboards accessible on the Web that allow managers to monitor operations and KPIs (key performance indicators) across the enterprise. Extending the information from the process remotely has become a viable way to extend the manufacturing footprint, consolidate expertise if it cannot be sourced locally and to manage asset performance more efficiently.

Smart Grid

Many energy companies are spending billions of dollars in developing smart grid to improve the efficiency, reliability, and sustainability of our electricity supply. Analysts project that the smart grid market will surpass $400 billion by 2020. The smartness in such a system is to allow the energy companies to collect data view demands at every level from large areas to individual homes. The goal is to use the data to control the flow of electricity when we really needed. This approach of supply per demand and price per demand can result in an overall saving on resources and reducing cost both on the part of the provider and the consumer.

UT (University of Texas), for example, is developing smart grid for the consumer, which will help the consumers to monitor their own electricity but also have the ability to return excess back to the Grid.

Video Surveillance

Video surveillance cameras and other sensors are key solution system to protect commercial building. The system software has capability to recognize faces, images, and count people and vehicles. This will contribute on increasing worker safety and lowering cost.

IoT and Mines

Sine Wave, a company that focuses on technological solutions for businesses, has created a customized IoT program that resulted in increased safety and communication in mines. According to their website, they designed a browser-based application that allows users to communicate with the workers, operators, and machines in the mine, as well as "see a real-time view of all activities underground [including] custom mapping of each mining operation." By knowing what is going on in the underground mines in real time, users can avoid safety hazards and respond to emergencies quickly.

NFC – Near Field Communications Technology

An NFC is a short-range secure technology that can be used by consumer to make a payment. For example, an NFC device can send data at a rate of 106, 212, or 424 kbits/s. NFC-enabled devices can be used in taxis, where payment can be done with swapping credit card but simply activating the app on your mobile device to start the payment process. This can expand to other adjacent markets such as restaurants and fast food, gas stations, food markets, and retail stores. Market introduction can be by introducing NFC as another method of payment adjacent to credit card payment devices. The overall benefit is to make the ecosystem cheaper and process of payment more convenient.

1.4.3.3 Infrastructure

Infrastructure for smart communities and cities is for sustainable environment and living, which include public transportation and highways, public safety, disaster management, smart education, and smart health care.

Effectively, a city is looking how to improve things by collecting information from sensors and analyze the data and improve its plan in appropriate areas.

Smart Cities

The effort by researchers to create human-to-human interface through technology in the late 1980s resulted in the creation of the ubiquitous computing discipline, whose objective is to embed technology into the background of everyday life. Currently, we are in the post-PC era where smartphones and other handheld devices are changing our environment by making it more interactive as well as informative.

A smart environment is the physical world that is richly and invisibly interwoven with sensors, actuators, displays, and computational elements, embedded seamlessly in the everyday objects of our lives, and connected through a continuous network.

The creation of the Internet has enabled individual devices to communicate with any other device in the world. The inter-networking reveals the potential of a seemingly endless amount of distributed computing resources and storage owned by various owners.

The advancements and convergence of MEMS technology, wireless communications, and digital electronics has resulted in the development of miniature devices having the ability to sense, compute, and communicate wirelessly in short distances. These miniature devices called nodes interconnect to form a WSN and find wide application in environmental monitoring, infrastructure monitoring, traffic monitoring, retail, and so on.

Waste Management
Waste management systems use IoT devices to monitor those who exceed waste limits. In some US cities, residential waste volume has declined because recycling measures were implemented. In the world, smart water systems and meters have reduced leaks and spillages with the help of sensors.

Large Event – Car Race
In a car race such as Formula-1, a race car can produce thousand's of data points every time it goes around the track. Designers and engineers to look for improvement as to how to get an extra mile or kilometer per hour can then analyze the data collected.

Cattle Wellness
Cattle: Researcher is working on recognizing and preempts illness or infection in cattle before affecting the rest of the herd.

This can be accomplished by placing wireless devices on cattle to monitor their behavior pattern including movement and sleep in real time. Such a system is expected to reduce the high cost of medication and help increase milk yield.

Crops (Vine Field) – Optimal Results
Where a variety of sensors can be placed in vineyard to measure real-time changes in environmental conditions such as air temperature, relative humidity, soil moisture, solar radiation, and leaf wetness, the system feeds the data into a cloud-based software platform that can be viewed by employees to respond in real-time event. The outcome of analyzing the data will help in harvesting at the optimal time and help in the management of resources such as water and fertilizer.

Smart Delivery System (Drone)
To improve on service delivery in particular when lightweight item is ordered, the desire to have immediate delivery as soon as the order is ready. This is where drone comes into play. The drone can pick up the item being ordered and via GPS can travel toward the destination, descent, send an alert to the customer before descending, drop the item at the front door of the customer, and return to the base. Services can be applied to food order, books, and others. The US FAA has started granting licenses for certain commercial applications. Debate is still on for safety and privacy. Amazon is about to launch such services. Amazon estimates 80% of its deliveries are light enough (<5 lbs) for a drone to carry. Today, customers and hobbyist can order and operate drones under certain guidelines such as flying height is 40 ft.

Property/casualty companies also are investigating the employment of sensors in conjunction with geolocation systems to build risk profiles across various properties.

Security and Privacy

Security has become and remains to be a key challenge to secure access of content at anytime and anywhere. Security risks is now associated with the rapidly proliferating IoT. For the first time, the Industry has reported actual proof of a cyber attack involving common appliances.

The "IoT" holds great promise for enabling control of all of the gadgets that we use on a daily basis. It also holds great promise for cybercriminals who can use our homes' routers, televisions, refrigerators, and other Internet-connected devices to launch large and distributed attacks. Internet-enabled devices represent an enormous threat because they are easy to penetrate, consumers have little incentive to make them more secure, the rapidly growing number of devices can send malicious content almost undetected, few vendors are taking steps to protect against this threat, and the existing security model simply will not work to solve the problem. Security strategy is required from the early stages of defining and developing an IoT solution looking at all aspects sensing, hardware, software, networking, cloud, storage, code, mobile devices, social media, processes, and others. Risks associated with access, data integrity, privacy, and network need to be identified, prioritized, and mitigated. Policy needs to be established as to handling the security risks. Technical requirement will include user identification and authentication, device identification and authentication, security function isolation, denial-of-service protection, and software and information integrity.

Privacy

All these intelligent devices and systems are generating a lot of data about individuals, homes, cars, industry, and cities. What happens to all the data collected? How is it protected? Data protection is becoming of increasing importance for both individuals and businesses. Various initiatives have started in this area but more innovative and collaborative work is needed!

Networking Infrastructure Solution

Networking Infrastructure Solution calls for providing reliable, secure operation in the most harsh environments and extreme temperatures to meet the most demanding applications, such as public safety, transportation, defense, oil and gas exploration, energy, and mining. Such solution exists in a scattered and silo fashion. More collaboration, integration is required to provide such secure network for operations.

1.4.3.4 Across Multiple Domains

IoT and Technologies

The IoT has many definitions, but revolves around connecting sensors, smart equipment, programmable logic controllers (PLCs), M2M, and RFID data with the Internet so that other systems or analytics software can respond to or make sense of the data. The great promise of the IoT is that information technology (IT) systems will have a real-time understanding of conditions, events, and material movements in the physical world.

M2M Opportunities

In the year 2020, the world will be more connected, wireless networks will connect more machines than people and M2M technology will help us to be more energy and cost-efficient, safer, and more secure. This is the era of connected intelligence.

The number of mobile connections is forecast by the GSM association to grow to 50 billion by the end of the decade. There will be several trillion wirelessly connected things, according to the Wireless World Research Forum. This increased use of mobile and M2M technology will deliver as much as $100 billion of annual savings thanks to material and energy efficiency savings (smart2020.org). The social and economic benefits are clear.

It is tempting to assume that cellular will hold the answer to M2M connectivity, but stop and ask yourself this question: are cellular networks designed for M2M or are they more about broadband connectivity for people?

It is worth noting that M2M devices communicate differently than humans and that there will be many more billions of them than us. These devices will communicate on a regular, optimized schedule and when they do speak, they do not say very much. So if machines and humans communicate so differently, why should machines and humans share the same networks? Perhaps one day that will be possible. Perhaps 5G cellular will deliver a common network approach. For the foreseeable future, M2M gateways will prevail in many scenarios.

M2M gateways offer the following key benefits:

Scaled to multiple settings (e.g., home, industrial, office, campus, city, and retail)

- A local intelligent node turns raw data into useful information
- A hub for cross-sector service and application convergence or "joined-up thinking"
- A secure node bridging broadband WAN and local area sensor networks, wireless and/or wired, which can even connect legacy-installed sensor/actuator nodes
- An aggregation node for a multitude of low-energy, low-cost sensor/actuator nodes.

1.4.4 C-IoT and RFID

1.4.4.1 Asset Management

Organizations are already using RFID tags combined with a mobile asset management solution to record and monitor the location of their assets, their current status, and whether they have been maintained.

1.4.4.2 Inventory Systems

An advanced automatic identification technology based on RFID technology has significant value for inventory systems. The system can provide accurate knowledge of the current inventory. The RFID can also help the company to ensure the security of the inventory. With the just in time tracking of inventory through RFID, the computer data can show whether the inventory stored in the warehouse is correct with quantity currently. Other benefits of using RFID include the reduction of labor costs, the simplification of business processes, and the reduction of inventory inaccuracies.

1.4.4.3 Product Tracking

RFID use in product tracking applications begins with plant-based production processes, and then extends into post-sales configuration management policies for large buyers.

1.4.4.4 Supply-Chair Merchandize Tracking

RFID can also be used for supply chain management in the fashion industry. The RFID label is attached to the garment at production can be read/traced throughout the entire supply chain and is removed at the POS.

1.4.4.5 Access Control (TX TAG, Conferences)

RFID tags are widely used in identification badges, replacing earlier magnetic stripe cards. These badges need only be held within a certain distance of the reader to authenticate the holder. Tags can also be placed on vehicles, which can be read at a distance, to allow entrance to controlled areas without having to stop the vehicle and present a card or enter an access code.

1.4.4.6 Transportation and Logistics

Logistics and transportation are major areas of implementation for RFID technology. Yard management, shipping and freight, and distribution centers use RFID tracking technology. In the railroad industry, RFID tags mounted on locomotives and rolling stock identify the owner, identification number, and type of equipment and its characteristics. This can be used with a database to identify the lading, origin, destination, and so on of the commodities being carried.

In commercial aviation, RFID technology is being incorporated to support maintenance on commercial aircraft. RFID tags are used to identify baggage and cargo at several airports and airlines. Some countries are using RFID technology for vehicle registration and enforcement. RFID can help detect and retrieve stolen cars.

1.4.4.7 Animal Identification

RFID tags for animals represent one of the oldest uses of RFID technology. Originally meant for large ranches and rough terrain, since the outbreak of mad-cow disease, RFID has become crucial in animal identification management. An implantable RFID tag or transponder can also be used for animal identification.

1.4.4.8 Sports

RFID for timing races began in the early 1990s with pigeon racing. RFID can provide race start and end timings for individuals in large races where it is impossible to get accurate stopwatch readings for every entrant.

1.4.4.9 Telemetry

Active RFID tags also have the potential to function as low-cost remote sensors that broadcast telemetry back to a base station. Applications of tagometry data could include sensing of road conditions by implanted beacons, weather reports, and noise level monitoring.

1.4.4.10 RFID – Big Data Filtering

Not every successful reading of a tag (observation) represents data useful for the purposes of the business. A large amount of data may be generated that is not useful for managing inventory or other applications. For example, a customer moving a product from one shelf to another, or a pallet load of articles that passes several readers while being moved in a warehouse, are events that do not produce data that is meaningful to an inventory control system.

Event filtering is required to reduce this data inflow to a meaningful depiction of moving goods passing a threshold. Various concepts have been designed, mainly offered as *middleware* performing the filtering from noisy and redundant raw data to significant processed data.

1.4.5 C-IoT and Nanotechnology

Nanotechnology will have impact on all aspects of IoT at the sensing level, gateway, and service levels. Here are some examples.

1.4.5.1 Food Safety

According to United Nations, 1.3 billion tons or one-third of global food production is wasted each year. In the future we expect nanotechnology will be embedded in our grocery and send alert as to when the food will go bad and thus ensure that the food on the shelf is always safe. Furthermore, typically the foods we eat are made of molecules of protein, fat, and carb. Research scientist is beginning to focus on nanotechnology to bring about super food that is good for our health.

1.4.5.2 Surgery

Nanotechnology can be used in some cases to replace surgery. For example, the small microsurgery material can be introduced through the vascular system and preprogrammed or guided by physician to perform various functions such as eliminated cancer cells before they can spread, changing new chromosome for old ones inside individual living human cells.

At the University of California, Nanotechnology can be used in developing particles that mimic the human cell called "nanosponges" that have a diameter of 85 nm or about 3000 times smaller than regular human blood cell. These nanosponges can be injected into the body by the millions, which can be used in various functions such as treating bacterial infections or eliminating foreign toxin by intercepting and attacking before damage to actual blood cell can be done.

1.4.5.3 Treatment

Nanotechnology can be used to have more effect than chemotherapy in attacking directly drug resistance cancer cells. Research has shown that nanotechnology treatment is 30 times more effective than traditional chemotherapy treatment and require 1/10 of chemical doze.

1.4.6 Cyber-Collaborative IoT (C^2-IoT)

This book provides a visionary approach transitioning from IoT (point solution) to the Future/Collaborative IoT (C-IoT solutions/systems). Moving forward toward Cyber-Collaborative IoT (C^2-IoT system of systems), we see major technology players such as nanotechnology, 5G to 10G, and others as disruptive technology that calls for cyber collaborative intelligence making sensing more intelligent, moving services from edge of the network to be distributed between End sensing node and cloud intelligence, inter-collaborative-cloud will remove global barriers and finally solutions and services will also be hosted by a distributed providers (as a result of consolidation among service providers, carriers, etc.). Inter-countries policies

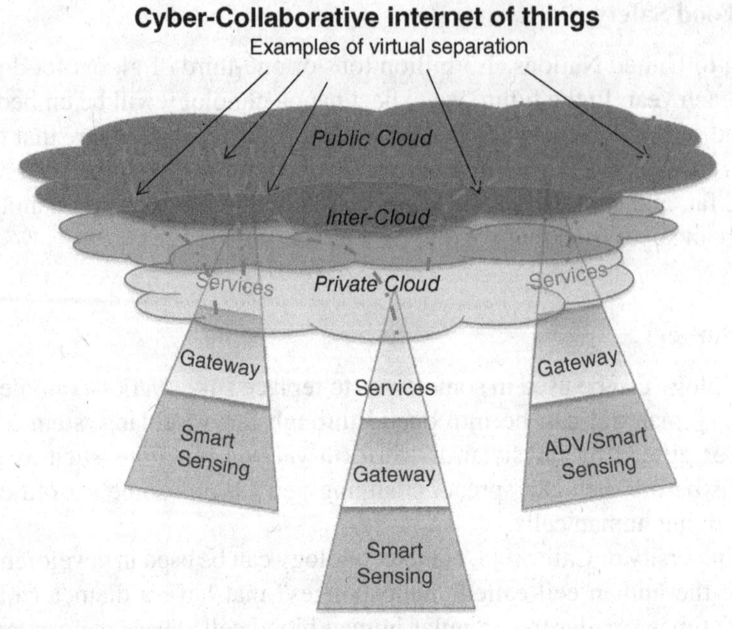

Figure 1.11 Future approach to scaling C-IoT

will be required to protect soft asset (information and intelligence). Privacy, safety will require a global level orchestration. Thus, the world (perhaps headed by the United Nations-IEEE and others?) can play a key role in fostering a new level of collaboration for advancing technology for humanity in terms of living conditions, peace, and economic/industrial growth.

Figure 1.11 exemplifies the future approach to scale IoT globally and across all domains with security safeguard against hackers, and ensuring privacy.

1.4.7 C^2-IoT and Ebola Case

Today, Ebola outbreak in West Africa has sickened over 8000 and killed over 4000. Because there is no off-the-shelf medicine to treat it, there is a high risk that Ebola will spread with people infected with the virus from Africa to other countries.

Ebola, being a very serious and contiguous virus, requires a global containment plan by the UN – World Health Organization. Such undertaking will require collaboration with many nations who possess advancement of medial technology and expertise in treatment and administration.

How will C^2-IoT help in saving lives in countries impacted by the virus and contain the spread of the virus to other regions and countries?

Outlined below is an example of a C^2-IoT-based plan and steps to be taken regarding the Ebola case. See Figure 1.12.

- Immediate creations of an Ebola Emergency Virtual Control Center to be staffed with experts from other nations to supervise treatment of those infected, and contain the virus from being spread locally, regionally, and globally.
- Establishing an emergency fund leveraging social media to get the message out for a worldwide donation of money that will help to pay for clinics, medicine, protective gear, instruments, devices, and others.
- Call for physicians and experts to collaborate, discuss cases and effect of the treatments on patients and possible approaches based on results from similar cases.
- Need of IoT wearable devices for infected patients to be remotely monitored minimizing chance on others not to be infected.
- For Treatment, Mobile Clinics will be installed to serve those that require treatment equipped with sensors and devices that are continually sending critical information to the cloud. Tracked information may include:
  - Critical information collected in real-time of each individual inside the clinic,
  - Treatment and effect on each patient,
  - Types of drugs entering the clinics,
  - Types of disposal of items leaving the clinics,
  - Individuals coming in or going out of the clinics.
- Deployment of Law and Enforcement officers who received special training to handle such situations to monitor clinics, labs, and individuals. Officers are to be equipped with advance monitoring devices with audio and video communications capabilities.
- Because of shortage in medicine, a more proactive smart manufacturing facility is to be established with focus on switching mode of operations from "time-to-market" to "time-to-service."
- Airports of infected regions of this virus are to screen all travelers leaving the countries, are to be accountable by responding to Ebola questionnaire.
- Key airports receiving travelers from infected regions are to be equipped with advanced screening equipment and protocol.
- Briefing: A Spokesperson of the *Ebola Emergency Virtual Control Center* is to provide regular briefings identifying issues and where help is needed.
- Umbrellas of interconnected websites dedicated to Ebola are to be created. Websites will be dedicated for each country being impacted, populated with information and Q&A that will be constantly updated. A mechanism is to be established to connect with families of those infected by the virus.
- Leverage social media to facilitate communications and generate awareness.
- Need for a group to prepare educational material and disseminate information and tips to media, schools, and public at large.

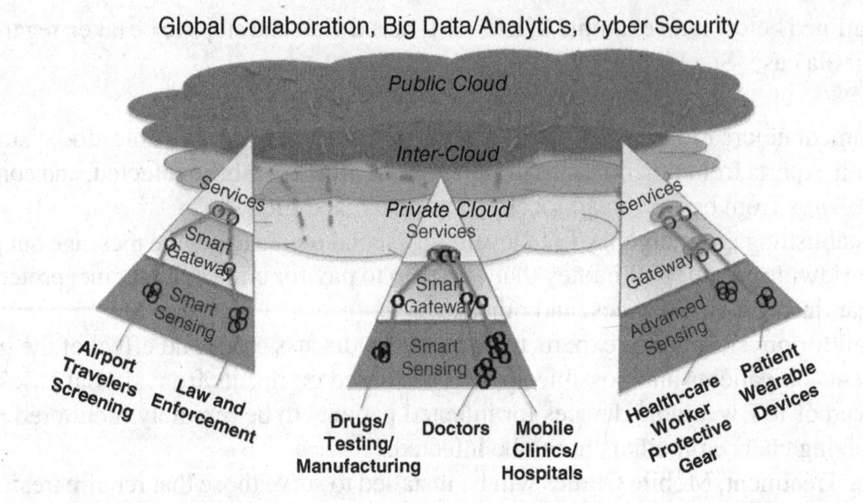

Figure 1.12 C^2-IoT model and Ebola case

- *Global Collaboration*. UN, leading nations, experts, medicine, fund, and so on.
- *Big Data*. Ebola drug R&D and usage, patient record, physician/workers records, mobile clinics/hospitals, and so on.
- *Analytics*. Demand for drug/physicians/clinics, effectiveness of medicine, education, and so on.
- *Cyber Security*. Law and Enforcement, protecting virtual with physical at local sites, other regions, cloud, and so on.

The above shows that C^2-IoT can be used without any physical boundaries connecting the physical (sensors, physicians, patients, clinics, drugs, etc.) with the virtual (information, knowledge, track history, trends, R&D results, insights, etc.). Such a system will have the major functions of sensing, connectivity to the clouds. The cloud will become a repository of all information pertaining to Ebola virus, patients, medicine, treatment, results, premises, and those who come in contact with patients. A great deal of computation and analysis of data collected will be required. Results and insights are to be shared among interested parties for the purpose of saving lives, treat those who are infected and protect the rest of the public from being impacted.

C^2-IoT will enable connecting people-to-people-to-systems-to-knowledge-to-insights protecting all for privacy, hackers, and breakdowns. Mobile smart devices such as smartphones and tablets with cognitive computational capabilities and natural voice language processing would be required. Devices such as Watson (won Jeopardy competition in 2011) could become such device that process information at a very high speed and assist experts with filtered information which may include specific data, trends, modeling, comparisons, and other critical information that are needed by those engaged in the networks as well as those in the *Ebola Emergency Virtual Control Center*.

1.4.8 Summary

In this section, we have witness examples of potential future application for C-IoT where vast amount of sensors enable connecting things to the Internet on a daily basis.

Future smart mobile/wearable device connected to the clouds (local, public, and inter-cloud) enabled by C-IoT can deliver a vast set of functionalities that can be personalized based on where you are and what function you are engaged in. Security will be embedded and privacy will be authenticated based on unique identification such as biometric signature.

As we witness today with Watson winning Jeopardy in 2011 exhibiting a capability of understanding natural language, navigate through a wealth of stored information and knowledge, and using sophisticated algorithms to filter through big data and retrieve most suitable response. Such approach is being prepared for many of the markets such as health, traffic, finance, agriculture, businesses and as such impacting Individual, Industry, and Infrastructure.

The future focuses not only in making available insightful information to individuals localized to where they are, and their needs at anytime.

A person may have multiple jobs, but try to balance his life focusing to what is important that will bring great experience and rewarding results.

With a mobile/wearable device, an individual controls the environment in the home (lighting, security, climate, wellness, etc.), reviews the day's schedule and missions to accomplish, receives assistance in the car to reach various destinations (car safety, traffic conditions, navigation, etc.), receives help and on-demand information for each of the missions (e.g., closest and cheapest gas station or electric power charge for the car and critical information and profile of the upcoming meeting), alerts and connects with business colleagues for lunch, prepares for an afternoon business meeting and gets profile of new individuals to meet, accesses to business records and stat related to the meeting, and is being alerted on the timing of the social function, and the device uses the smart parking function to direct for available parking slot. While the person is in a particular mission, he is constantly connected to the environment for any major alerts such as intruder approaching his home or entering his car – with a video surveillance he can in turn send the picture to the police with real-time information. Overall, this helps enrich experience, making time more efficient and generate rewarding results.

As mentioned such example can change scenario from time to time and relevant information and insights are readily available for viewing for action. Thus features and functions have been transported to a series of experiences of value being created. Thus, C-IoT business model is about creating experiences of value.

When the world around us becomes plugged in and aware, it will drive efficiencies like never before.

1.4.8.1 Collaborative IoT and Climate Change

While it may seem out of reach today (and possibly laughable to some), IoE will eventually allow us to become better stewards of our finite resources by improving how we sense, understand, and even manage our environment. As billions and even trillions of sensors are placed around the globe and in our atmosphere, we will gain the ability to literally hear our world's "heartbeat." Indeed, we will know when our planet is healthy or sick. With this intimate understanding, we can begin to eradicate some of our most pressing challenges, including hunger and ensuring the availability of drinkable water.

1.4.8.2 Secure, Reliable Connectivity in Extreme Environments

The rise of the IoT is bringing networking to new device types and locations. Networks must reliably and securely connect people and things in environments with weather and temperature extremes. These networks must also provide seamless mobile connectivity on devices with size, weight, and power limitations in any location, under any circumstance.

Wearable Devices, Health Monitoring, and Your TV

You are wearing a fitness/health device and watching television in your bedroom. The device is linked to your TV through shared logins or a simple linkage through your phone; so now, your TV knows about your body's activity levels including your sleep/wake patterns.

Advertisers will make smarter TV buys and stop sending commercials to sleeping consumers. Taking this one step further, since your fitness device knows your activity levels, it could inform which commercials you see – for example, more active consumers might get the stepper and workout equipment commercials which are wasted on me.

Wearable, Mobile, and TV

To build on the example above, looping your smart phone into this relationship will give advertisers an additional dataset. Your GPS and cell tower data tell us where you have been locally and if you have been out of town/on vacation. We know if you go to the gym, play tennis, or like to dine out.

Based on this specific data, we are now able to send you digital (smart phone) and analog (TV) ads that are highly relevant to your behavior (and travels).

Location-Based Understanding

How do advertisers know where you go? Our mobile devices are constantly collecting information about our location and movements.

Just check your iPhone under Settings (Privacy – Location Information – System Services – Frequent Locations). The iPhone by default records all your movements and patterns. Advertisers could leverage this data to better understand individual consumer behavior and determine their value more accurately.

For example, I go to a coffee shop once a week; my wife goes almost every weekday. In a real-time bidding environment, her value to a coffee advertiser should be five times greater than mine based on her location history.

Or we can use historic behavior to send the right ad at the right time to the right person. If we know that someone goes to the movies every Friday, we should start sending targeted movie ads to him on Thursday instead of Saturday.

Linking this back to the rise in beacons and GPS-enabled phones, it is conceivable that we will be able to measure the impact a billboard had on us based on visits to or engagements with brand assets while in proximity to the brand's billboard.

Web Analytics and Spatial Relations

I believe in 2015, we will be able to collect new forms of data, through the device browser, about how a mobile device is being used in its environment; this will allow us to segment web analytics or media consumption behavior in a much more granular way.

This next-generation information collection goes beyond device model and manufacturer to many more physical-world metrics such as device orientation in 3D space, ambient light levels, elevation of ground or change in user position (moving vs static).

We could then start to segment and target people based on a new set of options like whether someone is at home/work or on the move, or lying down as opposed to standing up.

Connected Homes

With connected refrigerators, lights, thermostats, and other sensors in the home, we can glean insights into the user's current environment. Is it dark? Is it warm or cold? Is the TV on or off? What are they watching? Who is home? This and more can be gained by connecting addressable advertising with connected homes.

Taking this one step further, sooner or later our fridge will know what we are low on, and send us ads or recommendations based on previous contents and consumption patterns. Our connected TVs will know what we watch and who is watching; this will allow marketers and networks to bring truly relevant content to the viewer's screen.

Fueled by the prevalence of devices enabled by open wireless technology such as Bluetooth, RFID, Wi-Fi, and telephonic data services as well as embedded sensor and actuator nodes, IoT has stepped out of its infancy and is on the verge of transforming the current static Internet into a fully integrated Future Internet [7]. Internet revolution led to the interconnection between people at an unprecedented scale and pace. The next revolution will be the interconnection between objects to create a smart environment.

References

[1] Behmann, F. (2014) How the Internet of Things Promotes Collaborative Innovation. Austin Business Journal (Oct. 20, 2014) http://www.bizjournals.com/austin/blog/techflash/2014/10/how-the-internet-of-things-promotes-innovation.html (accessed 10 December 2014).
[2] ABI https://www.abiresearch.com/press/iot-developers-to-total-3-million-in-2019-paving-t (accessed 18 November 2014).
[3] FDA – US Food and Drug Administration http://www.fda.gov (accessed 18 November 2014).
[4] Research2guidance http://www.research2guidance.com (accessed 18 November 2014).
[5] Gartner Research http://www.gartner.com/newsroom/id/2636073 (accessed 18 November 2014).
[6] Forbes http://www.forbes.com/sites/ptc/2014/07/01/how-the-internet-of-things-is-transforming-manufacturing/ (accessed 18 November 2014).
[7] Gubbi, J., Buyya, R., Marusic, S., and Palaniswam, M. (2012) Internet of Things (IoT): A Vision, Architectural Elements, and Future Directions.

2

Application Requirements

2.1 C-IoT Landscape

Advancement of technologies will continue to accelerate the evolution of IoT (Internet of Things), impacting many areas that will touch our lives.

In order to address technology and standards that are driving the collaborative Internet of Things (C-IoT), we introduce a C-IoT model that will be used as an illustration throughout the book. The model will be used to describe business apps requirements, define high-level architecture solution, and enabling technologies and protocols.

2.1.1 C-IoT Model and Architecture Layers

Figure 2.1 describes a generic C-IoT model consisting of three layers: Sensing, Gateway, and Services.

2.1.1.1 Sensing Layer

This layer enables interface to objects that are currently passive and where tapping into these objects will generate a stream of data and information that matter to IoT for enterprise for any given market and individual.

2.1.1.2 Gateway/Aggregation Layer

This layer enables the stream of data to move from one level to the next for additional processing. For example, this can be for moving from body area network (BAN), personal area network (PAN) to home area network (HAN) or from HAN to local area network (LAN) or from LAN to wide area network (WAN).

Collaborative Internet of Things (C-IoT): For Future Smart Connected Life and Business, First Edition.
Fawzi Behmann and Kwok Wu.

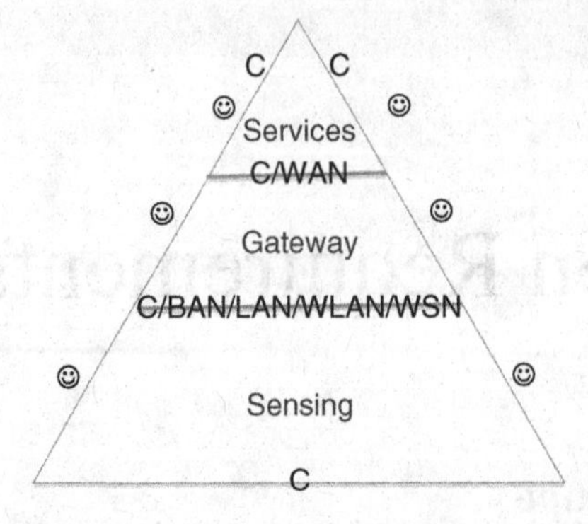

Figure 2.1 C-IoT model

2.1.1.3 Service Layer

This layer provides insights on the data collected from all layers and offers the insights as services to Individuals, Industries, or Infrastructures.

2.1.2 C-IoT Model and Enabling Technologies

Figure 2.2 describes and elaborates on C-IoT model with several attributes:

On the right-hand side of the triangle are Processing (MCU/MPU (microcontroller unit/microprocessor unit)), Connectivity (BAN/LAN/WLAN/WSN/WAN), storage (Server), and data analysis (Big Data/Analytics) in the three layers Sensing, Gateway, and Services.

On the left-hand side is the architecture view Physical, Virtual, and Cyber, highlighting enabling technologies and protocols.

Connectivity "C" exists throughout all the layers.

Several technologies and standards contribute to realizing building blocks architecture solution in these layers, enabling building smarter C-IoT solutions.

2.1.2.1 Sensing Layer

This layer enables interface to objects that are currently passive and where tapping into these objects will generate a stream of data and information that matter to IoT for enterprise for any given market and individual.

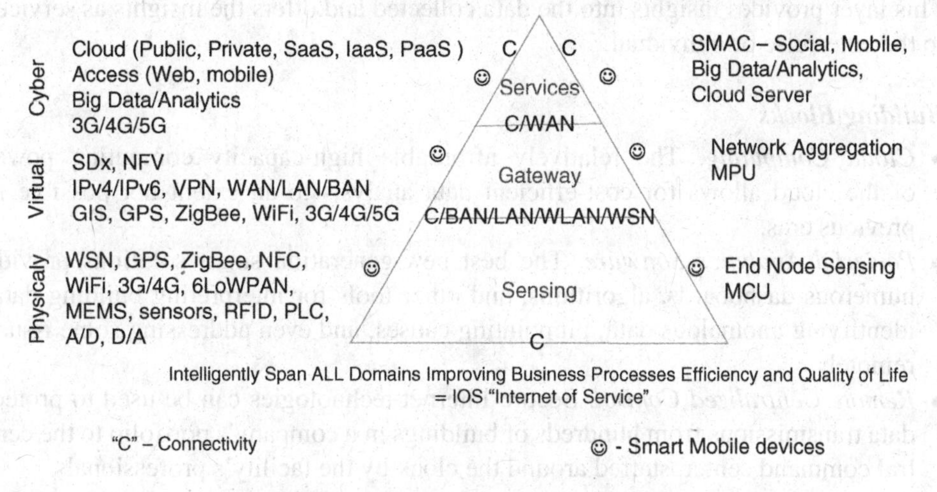

Figure 2.2 C-IoT elaborate model

Building Blocks

Wireless meters and sensors. Affordable wireless sensors and meters can now be used to monitor automated building equipment and relay data to a centralized remote command center.

2.1.2.2 Gateway/Aggregation Layer

This layer enables the stream of data to move from one level to the next for additional processing and services. Examples include BAN, PAN, LAN, and WAN.

Building Blocks

- *Internet*. The advent of the Internet and decreasing costs of data transmission now makes it financially feasible to transmit data from millions of building data points to the command center.
- *Open Data Communication Protocols*. Emerging networking standards such as Software Defined Networks (SDNs), Network Functions Virtualization (NFV), and SDx (Software Defined anything) for things like Storage, Data Center, and others are ways to help manage the configuration and operations of the network without the need to know the physical configuration. The trend is arriving to a platform that consists of products from different vendors that can interoperate. This enables moving away from vendor's proprietary protocol and places more emphasis on operating a heterogeneous network and support of cross-platform data sharing. Such a standardized and secure platform represents a significant milestone in the evolution of the IoT, and enables service providers to quickly and cost-effectively introduce differentiating IoT services.

2.1.2.3 Service Layer

This layer provides insights into the data collected and offers the insights as services to the enterprise or individual.

Building Blocks

- *Cloud Computing*. The relatively affordable high-capacity computing power of the cloud allows for cost-efficient data analysis to an extent not possible in previous eras.
- *Powerful Analytics Software.* The best new-generation smart solutions provide numerous dashboards, algorithms, and other tools for interpreting building data, identifying anomalous data, pinpointing causes, and even addressing some issues remotely.
- *Remote Centralized Control.* Secure Internet technologies can be used to protect data transmissions from hundreds of buildings in a company's portfolio to the central command center, staffed around the clock by the facility's professionals.

At each of the above layers, there are a set of protocols and standards that empower each layer. Following are descriptions of protocols and standards that impact the evolution of the C-IoT model.

2.1.3 Definition of Key Elements

Following is a list of terms and components associated with the C-IoT Architecture Model and specifically with the three layers Sensing, Gateway, and Services supported by illustrations and examples where appropriate.

2.1.3.1 Sensing Layer

Sensors

There are six sensors that we see embedded in many applications.

The sensor kit supports six common types of sensors:

- IR (infrared) temperature sensor
- Humidity sensor
- Pressure sensor
- Accelerometer
- Gyroscope
- Magnetometer.

The first three sensors measure environmental conditions important to practically all devices located in the field. The IR temperature sensor, for example, warns if a device's motor is overheating. A humidity sensor can detect if moisture is penetrating a waterproof casing and a pressure sensor can report on either excessive or substandard pressure. Taken together, these sensors can form powerful tools like a remote weather station.

The accelerometer and gyroscope are especially important to mobile instruments, as they allow a device's motion to be tracked independently of GPS (global positioning system) or other external location measurements. Most smartphones already have these features, but the sensor kit handles a wider range of conditions than most smartphone components. Finally, the magnetometer measures magnetic fields and electric currents, providing a safe means of remotely monitoring electric grids and power generators.

Microelectromechanical Systems

MEMS, standing for microelectromechanical systems, is a system with many small devices such as sensors, actuators, switches, robots, or other devices that can detect, for example, light, temperature, vibration, magnetism, or chemicals. They are usually operated wirelessly on a computer network and are distributed over some area to perform tasks, usually sensing through radio-frequency identification. Without an antenna of much greater size, the range of tiny smart dust communication devices is measured in a few millimeters and they may be vulnerable to electromagnetic disablement and destruction by microwave exposure.

The following are examples of applications/services:

- *Integrated Work-Order Management.* Today's building management systems can be integrated with a work-order system to streamline communications with on-the-ground facilities staff when human attention is required.

Radio-Frequency Identification

In the IoT paradigm, many of the objects that surround us will be on the network in one form or another. Radio-Frequency Identification (RFID) and sensor network technologies will rise to meet this new challenge, in which information and communication systems are invisibly embedded in the environment around us.

A radio-frequency identification system uses *tags* or *labels* attached to the objects to be identified. The tag can be a serial number, a license plate, or product-related information such as a stock number, lot or batch number, production date, or other specific information.

RFID systems can be classified by the type of tag and reader. A typical operation consists of an RFID reader that transmits an encoded radio signal to interrogate the tag. The RFID tag receives the message and then responds with its identification and other information.

RFID tags can be passive, active, or battery-assisted passive. An active tag has an on-board battery and periodically transmits its ID signal. A battery-assisted passive tag has a small battery on board and is activated when in the presence of an RFID reader.

Tags may either be read-only, having a factory-assigned serial number that is used as a key into a database, or may be read/write, where the system user can write object-specific data into the tag. Field programmable tags may be write-once, read-multiple; "blank" tags may be written with an electronic product code by the user.

RFID tags contain at least two parts: an integrated circuit for storing and processing information, modulating, and demodulating a radio-frequency (RF) signal, collecting DC (direct current) power from the incident reader signal, and carrying out other specialized functions; and an antenna for receiving and transmitting the signal. The tag information is stored in a nonvolatile memory. The RFID tag includes either a chip-wired logic or a programmed or programmable data processor for processing the transmission and sensor data, respectively.

Recently, decreased cost of equipment and tags, increased performance to a reliability of 99.9% and a stable international standard around UHF (ultra high frequency) passive RFID have led to a significant increase in RFID usage.

Global Positioning System

The GPS is a space-based satellite navigation system that provides location and time information. Commercial GPS software is available on various devices such as mobile phones and tablets. Although GPS-enabled smartphones are gaining ground in the portable navigation market, the standalone portable navigation device (PND) is far from dead. In fact, today's PNDs sport more features than ever to help you get from point A to point B quickly and safely – features such as audible driving directions with text-to-speech (TTS), spoken street names, real-time traffic updates, Internet connectivity for points-of-interest search, and large easy-to-read screens, to name a few.

White Space

Spectrum sharing and cognitive radio create new wireless services.

White Space is the name that has been given to unused TV channels in various locations around the United States and in some other countries. The unused channels can be deployed for other communications purposes, making free spectrum useful. Sometimes called TV White Spaces (TVWSs), these channels can be repurposed as needed. TWWS is called Super Wi-Fi and White-Fi as well.

Some standards for White Spaces have been developed, including IEEE (Institute of Electrical and Electronics Engineers) 802.11af, 802.22, and Weightless. The 802.11af is based on the existing Wi-Fi® standards but modified for the White Space bands. Wi-Fi, Bluetooth®, ZigBee®, cordless phones, microwave ovens, and many industrial products currently share the band from 2.4 to 2.483 GHz.

White Spaces are becoming an enabler for the IoT. The White Space spectrum could be used for connecting devices where infrastructure becomes intelligently interconnected allowing collected information to be passed between traditionally disconnected devices and hardware (referred to as machine-to-machine communication or M2M).

For example, cars could communicate with each other, warning drivers of stationary vehicles along their path that would otherwise not be visible due to traffic. Cars could also connect to the road infrastructure for traffic management, allowing intelligent adjustment of speed limits and traffic patterns to eliminate the stop–start traffic congestion often seen on motorways and ensuring a higher average speed and shorter average journey times.

Devices such as mobile phones and tablets could use the free spectrum by knowing which frequencies are available, at what power levels, and at which times of the day in a particular location.

Designed for IoT and White Space with the following considerations [1]:

- Deep indoor coverage with low transmit power
- Unlicensed operation bringing an interference risk that requires frequency hopping to mitigate
- Long battery life implying a sleep mode with periodic awakening (e.g., every 15 min)
- Base station processing to be moved to the core network
- The core network to be implemented in software running in the cloud.

Figure 2.3 shows IoT and White Space Flow Architecture.

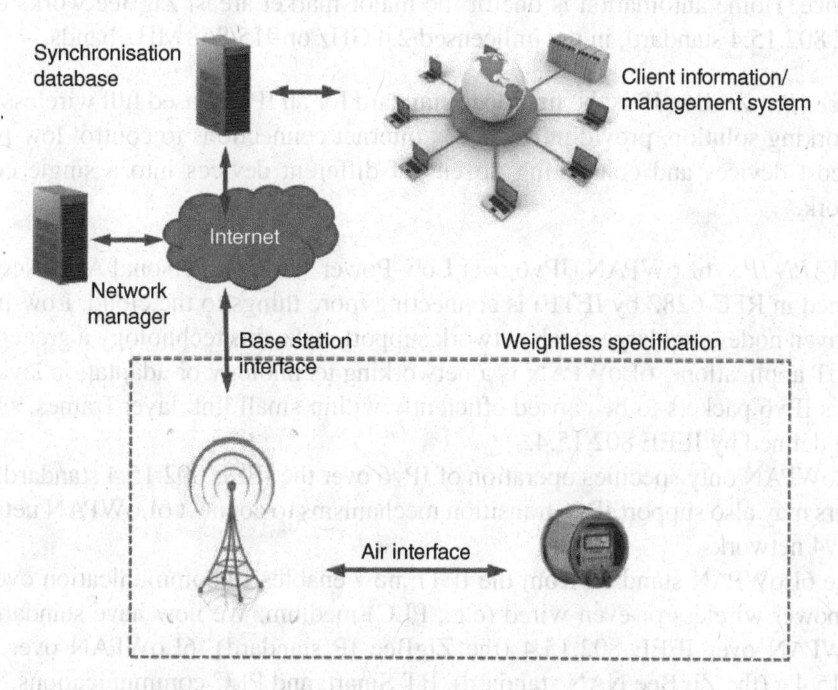

Figure 2.3 IoT and White Space flow architecture

IP Camera

An IP camera is the information source provider of the video security system based on IP network. It becomes more and more sophisticated with high-quality optics and also digital video accuracy.

Mainly three functions are important in an IP camera:

- *Pixel resolution.* Pixel resolution such as high-resolution 720p is already available. Full HD is coming in the near future.
- Compression rate andH.264 AVC is also mature enough now and SVC (scalable video coding) will come soon.
- Performance head room for Video Content Analysis.

A32-bit CPU (central processing unit) at 400 MHz is a minimum requirement now to perform any image processing in addition to the IP encapsulation.

Wireless/Cellular Network Connectivity

Bluetooth: Bluetooth is a short-range wireless with a range typically limited to 30 ft; it uses 2.4–2.48 GHz and a frequency hopping spread spectrum for transmission.

ZigBee: ZigBee is a low-power wireless communications technology designed for monitoring and control of devices and is maintained and published by the ZigBee Alliance. Home automation is one of the major market areas. ZigBee works on the IEEE 802.15.4 standard, in the unlicensed 2.4 GHz or 915/868 MHz bands.

ZigBee IP: ZigBee IP is the first open standard for an IPv6-based full wireless mesh networking solution, providing seamless Internet connections to control low-power, low-cost devices and connecting dozens of different devices into a single control network.

6LoWAPN IP: 6LoWPAN (IPv6 over Low-Power Wireless Personal Area Network) (defined in RFC 6282 by IETF) is connecting more things to the cloud. Low-power, IP-driven nodes, and large mesh network support make this technology a great option for IoT applications. 6LoWPAN is a networking technology or adaptation layer that allows IPv6 packets to be carried efficiently within small link layer frames, such as those defined by IEEE 802.15.4.

6LoWPAN only specifies operation of IPv6 over the IEEE 802.15.4 standard; edge routers may also support IPv6 transition mechanisms to connect 6LoWPAN networks to IPv4 networks.

The 6LoWPAN standard from the IETF now enables IP communication over any low-power wireless or even wired (e.g., PLC) medium. We now have standards for 6LoWPAN over IEEE 802.15.4 (the ZigBee IP standard), 6LoWPAN over IEEE 802.15.4 g (the ZigBee NAN standard), BT Smart, and PLC communications. These networks are typically designed for payloads under 127 bytes.

Other Considerations

ZigBee versus Z-Wave: According to the ZigBee Alliance, ZigBee Home Automation offers a global standard for interoperable products. Standardization enables smart homes that can control appliances, lighting, environment, energy management, and security as well as the expandability to connect with other ZigBee networks.

On the other hand, Z-Wave is described as a wireless RF-based communications technology designed for control and status reading applications in residential and light commercial environments. Target applications for Z-Wave are home entertainment, lighting and appliances control, HVAC (heating, ventilation, and air-conditioning) systems, and security.

Thus, both technologies address similar environments and applications. However, there are some differences both in the physical layer (PHY) and RF.

Z-Wave took the Sub-1 GHz approach, which has superior range versus the 2.4 GHz approach of ZigBee. However, Sub-1 GHz home automation requires different SKUs for different regions.

Z-Wave uses frequency-shift keying (FSK) modulation, which is good enough for the Sub-1 GHz environment. ZigBee is based on direct sequence spread spectrum (DSSS), which is a more advanced and robust modulation.

Protocol Aspects. Both ZigBee and Z-Wave support mesh network topology, which is a strong requirement toward the revolution of "IoT."

Body Area Network/Personal Area Network

A *BAN*, also referred to as a *wireless body area network*(*WBAN*)ora *body sensor network*(*BSN*),is a wireless network of wearable computing devices. BAN devices may be embedded inside the body, implants, surface-mounted on the body in a fixed position wearable technology or accompanied devices, which humans can carry in different positions. The development of WBAN technology started around 1995 around the idea of using wireless personal area network (WPAN) technologies to implement communications on, near, and around the human body. About 6 years later, the term "BAN" came to refer to systems where communication is entirely within, on, and in the immediate proximity of the human body [2].

Local Area

- 802.11b
- 802.11g
- 802.11a
- 802.11n
- 802.11ac

Spectrum: 2.4 and 5.8 GHz Unlicensed bands.
Channel Bandwidth: 20 MHz.

Modulation technologies:

- *CDMA*: 80211b at 2.4 GHz
- *OFDM*: 802.11a at 5.8 GHz, 802.11 g at 2.4 GHz.

Security is via station authentication.

Maximum range ˜100 M with clear LOS in LAN configuration. Some specialized point–point applications are up to 20 km.

Table 2.1 provides frequency and maximum data rate for 802.11x.

Wi-Fi Alliance is an organization of vendors and users that provides interoperability standards and testing to equipment compliant with IEEE 802.11 standards.

802.11ac increases the maximum data rate for a single client quite a bit compared to 802.11n. Most of the first 802.11ac access points use triple-stream MIMO (multiple input, multiple output), similar to today's top-end 802.11n access points, but have a maximum data rate of up to 1.3 Gbps. The increase comes from using 80 MHz channels and a new modulation scheme (256 QAM (quadrature amplitude modulation)). As the technology matures, the maximum data rate will further increase by taking advantage of even more MIMO streams, but that will not come for some time. See Figure 2.4 comparing data rates for different 802.11's.

Note that the above figure compares 802.11ac 80 MHz channels with 802.11n 40 MHz channels.

802.11ac will only work on the 5 GHz band. Nearly every wireless client supports the 2.4 GHz band, but unfortunately the band suffers from high interference levels and is quite crowded. In nearly all environments, the 5 GHz band does not suffer from as much interference or crowding as the 2.4 GHz band, and 5 GHz has more spectrum available for Wi-Fi channels. 802.11ac channels will be 80 MHz wide (compared to the 20 or 40 MHz channels of 802.11n), with the option to spread out to 160 MHz channels in the future, although at double the channel bandwidth compared to 80 MHz, there will only be half as many channels.

The speed, capacity, and performance improvements 802.11ac offers over 802.11n promise several compelling benefits to the enterprise. 802.11ac is well suited to handling streaming video, making it ideal for enterprises becoming increasingly reliant on video conferencing and collaboration and for enterprises that need to preserve service for streaming media and data-intensive applications. Additionally, as BYOD continues to grow its foothold in the enterprise, 802.11ac will help support the larger

Table 2.1 802.11x

	802.11b	802.11a	802.11g	802.11n	802.11ac
Frequency (GHz)	2.4	5	2.4	2.4/5	5
Max data rate	11 Mbps	54 Mbps	54 Mbps	450 Mbps	1.3 Gbps

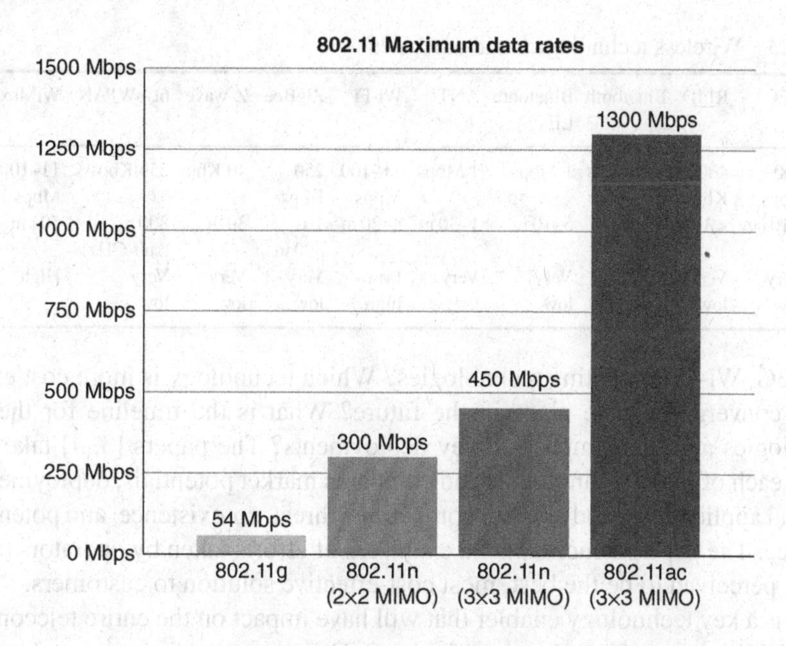

Figure 2.4 Comparisons of data rates among 802.11g,n,ac

numbers of devices connecting to corporate WLANs (wireless local area networks). Recent releases of consumer mobile devices, such as those running iOS and Android operating systems, offer 802.11ac compatibility to maximize performance.

Wide Area

- *2G/2.5G*
  - 2G: CDMA, TDMA, GSM
  - 2.5G EDGE, GPRS
- *3G* – CDMA2000, W-CDMA, UMTS, HSDPA
- *4G* – LTE and WiMax

Table 2.2 illustrates the up/down link rates for different cellular generations.

More wireless specifications about communication technologies are provided in Table 2.3.

Table 2.2 Cellular DL/UL rates

	GPRS	EDGE	UMTS	HSPA
DL	80 kbps	237 kbps	384 kbps	Up to 7.3 mbps
UL	20 kbps	59 kbps	384 kbps	Up to 5.4 mbps

Table 2.3 Wireless technologies and attributes

	NFC	RFID	Bluetooth	Bluetooth LE	ANT	Wi-Fi	ZigBee	Z-wave	6LoWPAN	WiMax	2.5–3.5 G
Speed	400 Kbps	400 Kbps	700 Kbps	1 Mbps	1 Mbps	11–100 Mbps	250 Kbps	40 Kbps	250 Kbps	11–100 Mbps	1.8–7.2 Mbps
Range	<10 cm	<3 m	<30 m	5–10 m	1–30 m	4–20 m	10–300 m	30 m	800 m (sub-GHz)	50 km	Cellular net
Power	Very low	Very low	Low	Very low	Very low	Low–high	Very low	Very low	Very low	High	High

Are 3G, Wi-Fi competing technologies? Which technology is more cost effective? Will a convergence take place in the future? What is the timeline for these three technologies and their impact on key deployments? The papers [3, 4] take a closer look at each of these technologies and compares market potentials, deployment costs, potential applications, and areas of competitive threat, co-existence, and potential convergence. The papers conclude with examples of efforts taken by operators to deliver what is perceived to be the best, most cost-effective solution to customers.

LTE is a key technology enabler that will have impact on the entire telecom supply chain. LTE impact affects semiconductor SoC, communications networking infrastructure, mobile devices, applications, and quality services transforming means of communications to a new level – higher speed, multimedia content, and enriching personal experiences. The paper [5] covers a number of important topics covering the needs for LTE, LTE market positioning and benefits, LTE market trends, deployment and applications, and LTE roadmap. The paper concludes with an example of enabling differentiated solution for LTE.

SOC architecture is an important building block of wireless solution. Advancement in SoC architecture has bridged the gap between semiconductor vendors, equipment vendors, and service providers. The presentation will address evolution in SoC in the past decade focusing on multicore, security, and power management as key SOC architecture factors for a scalable SOC platform. The paper [6] provides an example of base station on chip that supports multiple standards including LTE, integrated powerful MPU, DSP, and accelerator cores in a heterogeneous, many-core implementation.

2.1.3.2 Gateway Layer

IPv4/IPv6

With increase in the number of enabling devices to connect to the Internet, IP address range need to be increased from IPv4 to IPv6. For example, smart meters can now have an IP address. This in turn will enable new applications such as IPv6 used in smart grid enabling smart meters and other devices to build a micro mesh network before sending the data back to the billing system using the IPv6 backbone. Other examples can be (but not limited to) automation and entertainment applications in home, office, and factory environments. The header compression mechanisms standardized in RFC6282 can be used to provide header compression of IPv6 packets over such networks.

Semiconductor Moore's Law and IoT

The International Technology for Semiconductor Roadmap (ITRS) has assumed the validity of Moore's law traced back to a paper by Gordon Moore in 1965. Since 1970, the number of components per chip has doubled every 2 years. This historical trend has become known as "Moore's Law." As a consequence of this trend, the miniaturization of circuits by scaling down the transistor has dramatically decreased the cost per elementary function (e.g., cost per bit for memory devices, or cost per MIPS (million instructions per second) for computing devices); this has been the principal driver for the semiconductor technology roadmap for more than 40 years.

The industry is now faced with the increasing importance of a new trend, the need to combined digital and nondigital functionalities in an integrated system. This is translated as a dual trend in the ITRS: miniaturization of the digital functions ("More Moore") and functional diversification ("More than Moore") introduced in 2007 ITRS publications. "More Moore" continues with geometric scaling to improve density, performance and reliability, and equivalent scaling, including novel design techniques and technology such as multicore design as a continuation of "Moore's Law." "More than Moore" allows for the nondigital functionalities (e.g., RF communication, power control passive components, sensors, and actuators). Thus, the new guidance from ITRS becomes more relevant to IoT and understanding processor, system trends, and impact on industry and applications [7]. Figure 2.5 shows Moore's Law and More.

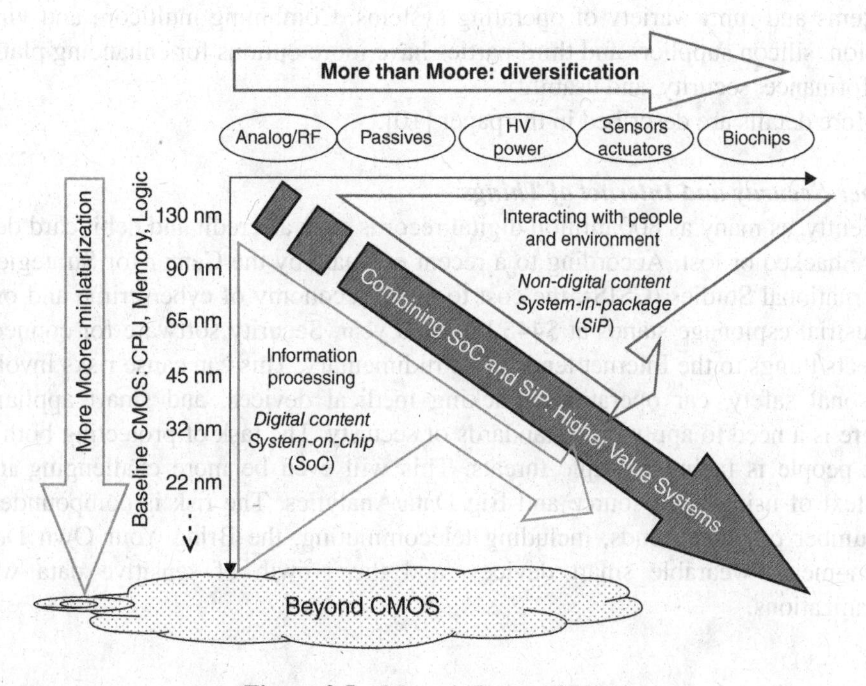

Figure 2.5 Moore's Law and More

Continuous miniaturization enables the integration of increased number of transistors on a single chip – system-on-chip (SoC). In 2007, the ITRS System Driver working group identified the SOC driver to include MPU, embedded memory, accelerators, I/O, floating points rise above the MPU driver. The networking driver, which includes SoC and software that target the embedded networking space, represents the rich integration of both "More Moore" and "More than Moore." A brief description is provided in the ITRS 2007 – System Driver publication [8]. More details can be found in a white paper written on the subject [9]. Future work can align more IoT to SoC and become part of the ITRS System Driver evolution.

Semiconductor SoC and Virtualization

The trend continues toward more powerful embedded systems with multicore processors and SoCs. Virtualization presents opportunities to reduce hardware costs and power consumption while enabling new platform-level capabilities. It allows a device to run multiple operating environments and share the underlying processing cores, memory, and other hardware resources.

The term "virtualization" is overused and needs to be broken down to hardware and software to clearly see which virtualization technologies are most relevant to embedded systems developers.

Hardware virtualization provides a more efficient mechanism for partition/OS switching and hardware resource allocation for a software-virtualized environment. Software virtualization takes advantage of hardware mechanisms to provide a platform, which enables embedded system designers to flexibly partition their systems and run a variety of operating systems. Combining multicore and virtualization, silicon suppliers and third parties have more options for enhancing platform performance, security, and usability.

More details are described in the paper [10].

Cyber-Security and Internet of Things

Recently, as many as 800 million digital records such as credit and debit card details were hacked or lost. According to a recent estimate by the Center for Strategic and International Studies (CSIS), the cost to global economy of cyber-crime and online industrial espionage stands at $445 billion a year. Security software for connecting objects/things to the Internet tends to be rudimentary. This can cause risks involving personal safety, car operations, hacking medical devices, and smart appliances. There is a need to apply high standards of security. The task of protecting both data and people is facing multiple threats. This will even be more challenging in the context of using open source and Big Data/Analytics. The risk is compounded by a number of other trends, including telecommuting, the Bring Your Own Device phenomena, wearable smart devices, and the growth of sensitive data within organizations.

Security Standards for Wi-Fi

WEP	Provides weak security Requires manual key management
WPA	Provides dynamically generated keys Provides robust security for small networks
WPA2	Requires manual management of pre-shared key Provides robust security for small networks
802.1X	Requires configured RADIUS server Provides dynamically generated keys

Software Defined Network (SDN)/Network Functions Virtualization

Over the past 5 years or so, it was easy for developers to embrace the cloud without dealing with the plumbing and wiring IT typically deals with. For IT to embrace the cloud, they need a more agile, enabling technology platform. SDN becomes a gateway to that reliable off-premises solution. Further, addition of security and other application services in the Layer 4–7 realm – by companies such as F5 – will deliver a holistic solution that works like a cloud environment but can be wired up in a way that IT can understand, trust, and embrace.

The new realities of a maturing SDN – combined with software-defined application services – will change the perception of the possible and result in enterprise IT solutions that look a lot more like consumer-oriented Web properties such as Netflix, Etsy, and Digg.

A big factor is SDN, which is going to be a key, underpinning element to this shift. The plumbing that runs beneath application and layer 4–7 architectures needs to be agile enough to support the elasticity and structure of the cloud as a run-time environment. While L4–7 technologies and the cloud have afforded a more fabric-based environment aligned with what the cloud promises, core switching has been a bit behind.

SDN is a new architecture that has been designed to enable more agile and cost-effective networks. The Open Networking Foundation (ONF) is taking the lead in SDN standardization, and has defined an SDN architecture model as depicted in Figure 2.6. SDN has three layers: Infrastructure layer (physical layer), Control Layer (Logical Layer), and Application Layer [11].

The key benefit of SDN is to automate network provisioning, monitoring, and control. This will include automate traffic monitoring and control over Layers 2 and 3, secure flow update, and L2/L3 GRE over IPSEC.

The Network Functions Virtualization concern is to automate distribution of network services across network appliances as virtual machines (VMs) that provide isolation (safety) across VMs. Examples of distributing L4–7 network services include Firewall, VPN (virtual private network)/IPSec, Deep Packet Inspection

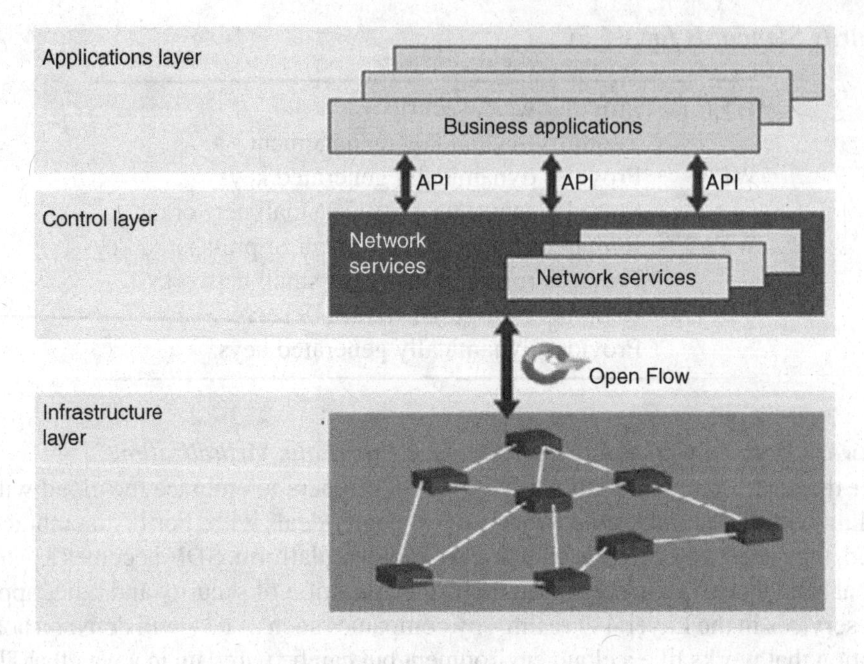

Figure 2.6 ONF/SDN architecture

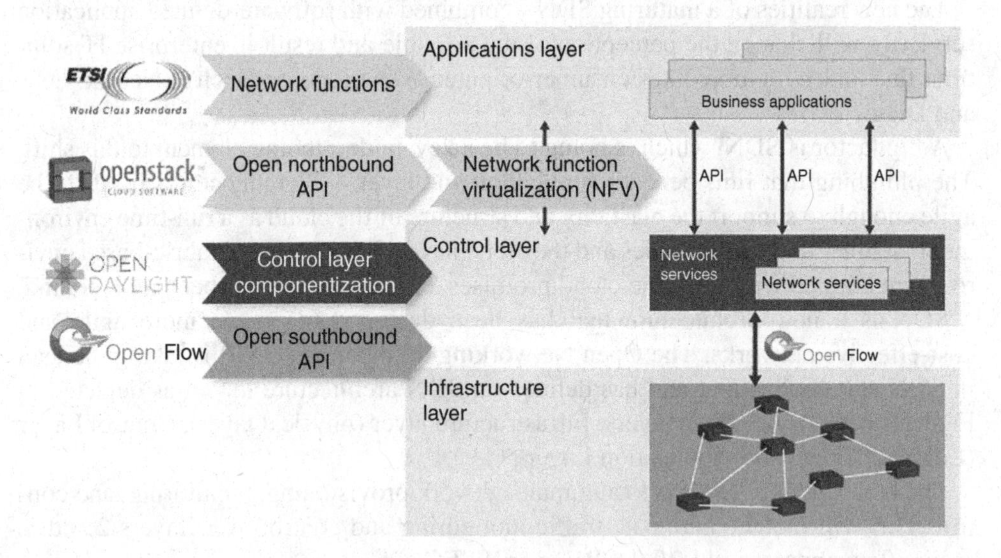

Figure 2.7 NFV (open stack + OVS)

(DPI), and Intrusion Detection system/intrusion prevention system (IDS/IPS). Automated VM migration can be deployed for load balancing and improving resource sharing. Figure 2.7 shows NFV positioned at higher level than SDN and focus on handling of network services across network appliances.

Geographic Information System

A geographic information system (GIS) is a computer system designed to capture, store, manipulate, analyze, manage, and present all types of geographical data. GISs have transformed the way spatial (geographic) data, relationships, and patterns in the world are able to be interactively queried, processed, analyzed, mapped, modeled, visualized, and displayed for an increasingly large range of users, for a multitude of purposes.

2.1.3.3 Services Layer

Embedded design is becoming more complex as design parameters evolve and demands for quicker time-to-market increase, especially for emerging IoT applications. Working closely with embedded software platform providers enhances the ability to provide the flexibility, reliability, scalability, and ease of use desired by customers developing the next generation of the IoT and graphical user interfaces (GUIs) for a high-performance, connected MCU.

MCU-based embedded solutions offer developers a software strategy based on the "application platform" approach. The approach contributes to expand product ranges, to accelerate specification process, and to improve user experience for markets such as smart grid, smart metering, smart appliances, building and home automation, and other IoT applications.

Embedded Service Switches can collect data from remote locations for real-time analysis by experts located at a remote site or headquarters. They can also help monitor the transit of vehicles and equipment to remote sites. Connected sensors can monitor the condition of the vehicles, cargo or crew, and convey this information to remote data centers or to other vehicles.

Industries are seeing unprecedented levels of automation and supply chain efficiencies as industrial control systems (ICSs) connect to the Internet. The IoT is certain to bring even greater acceleration of connectivity, not only in the production process and supply chain, but throughout all business processes. Businesses that respond to these innovations and move toward improved interconnectivity can become more globally competitive and ultimately lead in their markets.

Big Data

Today, big data is not limited to traditional data warehouse situations, but includes real-time or line-of-business data stores used as the primary data foundation for online applications that power key external or internal business systems.

It used to be that these transactional/real-time databases were typically "pruned" so they could be manageable from a data volume standpoint. Their most recent or "hot" data stayed in the database, and older information was archived to a data warehouse via extract–transform–load (ETL) routines.

But big data has changed dramatically. The evolution of the Web has redefined the following:

- The speed at which information flows into these primary online systems
- The number of customers a company must deal with
- The acceptable interval between the time that the data first enters a system and its time of exit
- Transformation of the data into information that can be analyzed to make key business decisions
- The kind of data that needs to be handled and tracked.

Some analysts such as Gartner have attempted to categorize these changes by describing big data as follows:

1. *Velocity.* How fast the data are coming in
2. *Variety.* All types of data are now being captured (structured, semi-structured, unstructured)
3. *Volume.* Potential of terabytes to petabytes of data
4. *Complexity.* Involves everything from moving operational data to big data platforms
5. Difficulty in managing the data across multiple sites and geographies.

Metcalfe's Law

Coined by Robert Metcalfe, the inventor of Ethernet, *Metcalfe's Law* states that *the value of a network grows by the square of the size of the network.* The idea behind this law is that a network's value is increased as the size of the network increases; this law is often referred to when talking about the Internet's value. For example, if the network has five machines, its value would be 25 ($5^2 = 25$), but if another network had 1000 machines its value would be 1 000 000.

This law is also considered applicable to more than just the Internet or a computer network. For example, a software product may increase in value as it grows in size. If a product only has 25 users, it is less likely to be known and used. However, if the same product has 1000 users, it is more likely to be known and used [12].

As more things, people, and data become connected, the power of the Internet (essentially a network of networks) grows exponentially. By combining people, process, data, and things, the exponential power of the Internet will allow us to create exponential responses to the extraordinary challenges faced by individuals, businesses, and countries.

2.1.4 Requirement Considerations

We present a taxonomy that will aid in defining the components required for IoT from a high-level perspective. There are three IoT components, which enable seamless

1. *Sensing*: made up of sensors, actuators, and embedded communication hardware
2. *Gateways*: connectivity and aggregation and virtualization

3. *Cloud/analytics* for on-demand storage and computing tools for data analytics and application/services.

2.1.4.1 Sensing

Radio-Frequency Identification (RFID)

RFID technology is a major breakthrough in the embedded communication paradigm, which enables design of microchips for wireless data communication. The microchips help in automatic identification of anything they are attached to, acting as an electronic barcode. The passive RFID tags are not battery powered and they use the power of the reader's interrogation signal to communicate the ID to the RFID reader. This has resulted in many applications particularly in retail and supply chain management. The applications can be found in transportation (replacement of tickets, registration stickers) and access control applications as well. The passive tags are currently being used in many bankcards and road toll tags, which are among the first global deployments. Active RFID readers have their own battery supply and can instantiate the communication. Of the several applications, the main application of active RFID tags is in port containers for monitoring cargo.

Sensing – Wireless Sensor Networks (WSN)

Recent technological advances in low-power integrated circuits and wireless communications have made available efficient, low-cost, low-power miniature devices for use in remote sensing applications. The combination of these factors has improved the viability of utilizing a sensor network consisting of a large number of intelligent sensors, enabling the collection, processing, analysis, and dissemination of valuable information, gathered in a variety of environments [13]. Active RFID is nearly the same as the lower-end WSN (wireless sensor network) nodes with limited processing capability and storage. The scientific challenges that must be overcome in order to realize the enormous potential of WSNs are substantial and multidisciplinary in nature. Sensor data are shared among sensor nodes and sent to a distributed or centralized system for analytics. The components that make up the WSN monitoring network include

1. *WSN Hardware*. Typically, a node (WSN core hardware) contains sensor interfaces, processing units, transceiver units, and power supply. Almost always, they comprise of multiple A/D converters for sensor interfacing and more modern sensor nodes have the ability to communicate using one frequency band making them more versatile.
2. *WSN Communication Stack*. The nodes are expected to be deployed in an ad hoc manner for most applications. In designing an appropriate topology, routing and MAC layer are critical for scalability and longevity of the deployed network. Nodes in a WSN need to communicate among themselves to transmit data in single or multi-hops to a base station. Node dropouts, and consequent degraded network lifetimes, are frequent. The communication stack at the sink node should be able to interact with the outside world through the Internet to act as a gateway to the WSN subnet and the Internet.

3. *Middleware*. This is a mechanism to combine cyber infrastructure with a Service Oriented Architecture (SOA) and sensor networks to provide access to heterogeneous sensor resources in a deployment-independent manner. This is based on the idea of isolating resources that can be used by several applications. A platform-independent middleware for developing sensor applications is required, such as an Open Sensor Web Architecture (OSWA). The Open Geospatial Consortium (OGC) builds OSWA upon a uniform set of operations and standard data representations as defined in the Sensor Web Enablement Method (SWE).
4. *Secure Data Aggregation*. An efficient and secure data aggregation method is required for extending the lifetime of the network as well as ensuring reliable data collected from sensors. Node failure being a common characteristic of WSNs, the network topology should have the capability to heal itself. Ensuring security is critical as the system is automatically linked to actuators and protecting the systems from intruders becomes very important.

2.1.4.2 Gateways

Addressing Schemes

The ability to uniquely identify "Things" is critical for the success of IoT. This will not only allow us to uniquely identify billions of devices but also to control remote devices through the Internet. The few most critical features of creating a unique address are uniqueness, reliability, persistence, and scalability.

Every element that is already connected and those that are going to be connected must be identified by their unique identification, location, and functionalities. The current IPv4 may support to an extent where a group of cohabiting sensor devices can be identified geographically, but not individually. The Internet Mobility attributes in the IPV6 may alleviate some of the device identification problems; however, the heterogeneous nature of wireless nodes, variable data types, concurrent operations, and confluence of data from devices exacerbates the problem further.

Persistent network functioning to channel the data traffic ubiquitously and relentlessly is another aspect of IoT. Although, the TCP/IP takes care of this mechanism by routing in a more reliable and efficient way, from source to destination, the IoT faces a bottleneck at the interface between the gateway and wireless sensor devices. Furthermore, the scalability of the device address of the existing network must be sustainable. The addition of networks and devices must not hamper the performance of the network, the functioning of the devices, the reliability of the data over the network, or the effective use of the devices from the user interface.

Gateways – Visualization

Visualization is critical for an IoT application as this allows interaction of the user with the environment. With recent advances in touch screen technologies, use of smart tablets and phones has become very intuitive. For a layperson to fully benefit from the IoT revolution, attractive and easy to understand visualization have to be created. As we move from 2D to 3D screens, more information can be provided to the user in meaningful ways for consumers. This will also enable policy makers to convert data into knowledge, which is critical in fast decision making. Extraction of meaningful information from raw data is nontrivial. This encompasses both event detection and visualization of the associated raw and modeled data, with information represented according to the needs of the end user.

2.1.4.3 Services

Cloud/Analytics

One of the most important outcomes of this emerging field is the creation of an unprecedented amount of data. Storage, ownership, and expiry of the data become critical issues. The Internet consumes up to 5% of the total energy generated today and with these types of demands, it is sure to go up even further. Hence, data centers that run on harvested energy and are centralized will ensure energy efficiency as well as reliability. The data have to be stored and used intelligently for smart monitoring and actuation. It is important to develop artificial intelligence algorithms, which could be centralized or distributed based on the need. Novel fusion algorithms need to be developed to make sense of the data collected. State-of-the-art nonlinear, temporal machine learning methods based on evolutionary algorithms, genetic algorithms, neural networks, and other artificial intelligence techniques are necessary to achieve automated decision-making. These systems show characteristics such as interoperability, integration, and adaptive communications. They also have a modular architecture both in terms of hardware system design as well as software development and are usually very well suited for C-IoT applications.

2.1.5 C-IoT System Solution – Requirement Considerations

A C-IoT solution is designed to run on classes of hardware devices that are severely constrained in terms of memory, power, processing power, and communication bandwidth. A typical C-IoT solution system has memory on the order of kilobytes, a power budget on the order of milliwatts, processing speed measured in megahertz, and communication bandwidth on the order of hundreds of kilobits/second. This class of systems includes both various types of embedded systems as well as a number of old 8-bit computers.

Understanding power efficiency is also critical as devices (such as wearable devices) are often not connected to power supplies and have to operate using energy harvesting sources or a single battery for several years without maintenance or battery replacement.

In addition to power consumption, connected device developers must consider factors such as system cost, component count, MCU performance, system size, standards, interoperability, security, ease of use, and in-field troubleshooting.

Adding wireless connectivity to remote devices not easily reached by Ethernet cable or power-line communications is another IoT design challenge that can be addressed by embedded developers with RF expertise.

Finally, software is required to bridge connected devices, aggregate sensor data, and present information to end users in an intuitive way via displays or over the Internet to their computers, tablets, or smartphones.

This chapter will focus on establishing IoT networking and architectural requirements utilizing the reference model introduced in the previous chapter.

The requirements will translate into arriving to an IoT energy-efficient device that may include

- Smart Sensing
- MCU platform with built-in wireless capabilities
- Support of a range of operating systems
- Development platform making the device available to entrepreneurs, who can build their own products on top of the design.

2.1.5.1 Smart Sensors

The trend toward a more useful world of connected devices hinges on a new class of smart sensors. Sensor integration and sensor fusion will continue to be the watchwords for these new sensor networks. In turn, more sophisticated sensor systems will rely on a new class of processors – low power and high performance data lower price point needed to enable widespread use. We may expect more everyday devices to become intelligent and networked.

2.1.5.2 Sensor Fusion

From Transducer to System on a Chip

Sensing technology has advanced far from the single-point transducer and bias circuitry. Modern sensor systems can incorporate bias, temperature compensation; signal conditioning, and even filtrations and A/D conversion. What is more, serial-bus multichannel sensor systems on a chip take advantage of fairly high-speed serial protocols such as IIC (Industrial Internet Consortium) and SPI. This reduces the cabling requirements and lets sensor circuitry remain close to what it is sensing to reduce noise.

While more sensors are being integrated into microcontrollers to form smart sensor nodes, multiple sensors are still delivering sensing data that need to be processed and analyzed to result in decision-making and trigger actions.

Sensor fusion is taking hard sensor data and fused with context-aware data such as location and owner data referred to as *soft data* (who, what, when, where, etc.) to perform situation assessment analysis, following which rule-based/knowledge-based decision-making actions can be taken intelligently at the edge nodes. The C-IoT service platform, which leverages sensor fusion software framework components, will perform ubiquitous sensing and processing works transparent to the user. This will enable the development of C-IoT applications that interoperate among multiple point solutions of Smart things. This will further drive the IoT market products to new heights.

2.1.5.3 Robotic and Sensor Fusion – Requirements

Most modern robots, such as room vacuum cleaner robots and pool cleaners, perform preprogrammed tasks and routines. Sensors are no longer an obstacle when designing a robotic control loop, and can be an integral part of the actuator/control mechanism. Clearly defined sensor types, limits, controls, and safeties are relatively easy to comprehend. But as evolved robots are more able to do complex tasks, the interleaving of sensor data to be used for real-time decision-making becomes more crucial. The fusion of sensory data is where the big picture is looked at rather than an individual control loop.

The requirement for more-capable and less-expensive robots calls for developing real-time control loops for a variety of functions including manipulating spinning, rotation of parts, vibration, and temperature.

2.1.5.4 MCU and C-IoT

The microcontroller is a genesis of the IoT. Over the past few decades, MCUs have been increased in functionality by embedding intelligence into the electronics around them. We are in an era where devices will connect with people and other devices.

2.1.5.5 MCU Low-Power Operation

IoT systems are severely power-constrained. Battery-operated wireless sensors may need to provide years of unattended operation, with little means to recharge or replace their batteries. An IoT system provides a set of mechanisms for reducing the power consumption of the system on which it runs.

The following features characterize the MCUs:

1. *Low Power.* Need very small operating current (32–80 MHz, NOP instructions). Reduction in power consumption can be further achieved by implementing a low-power "Snooze" mode in addition to the three traditional power management modes (Run, Halt, and Stop). The "Snooze Mode" allows common peripherals (i.e., ADC (analog-to-digital converter) or UART) to operate independently, while CPU is disabled. In this mode, power dissipation can be reduced by over 30% in comparison to an implementation without this mode. In "Halt Mode," the CPU is disabled, but all peripheral functions are operable. In this mode, with running RTC

(real-time clock) and LVD (low-voltage detection), power dissipation is reduced to ˜0.5 μA/MHz. In "Stop Mode," the high-speed clock oscillator and internal high-speed oscillator are disabled, so lowest power consumption is possible. For example, with running WDT (watch dog timer) and LVD, the current is reduced to about 0.5 μA/MHz.

2. *Scalability*. Offer a wide range of devices, which are available at a wide range of pin packages and different amount of Flash.
3. *High Efficiency*. – Have high-performance DMIPS (Dhrystone Million Instructions per Second)/MHz. This also calls for a maximum number of instructions to be executed in one clock cycle. This may call for implementing DMA/DTC/ELC functionality. For example, ELC reduces interrupt processing and improves real-time performance.
4. *Cost-Saving Features at System Level.* A high-precision (±1%) on-chip oscillator (32/64 MHz) makes 32 MHz CPU operation possible, without external oscillators and built-in features such as a reset circuit, LVD, WDT, and data flash with background operation function reduce the system cost.
5. *Security Features*. Safety features in hardware are required to ensure compliance with IEC/UL (International Electrotechnical Commission) 60 730. Examples include hardware WDT, TRAP instruction, illegal memory access, and frequency detection, Flash memory CRC, RAM parity error detector, RAM and SFR guarding, and A/D converter tests.
6. *An Extensive Ecosystem*. Offers developers to have access to development tools and kits, third-party network, online resources, engineering community, and online training.

2.1.5.6 Networking IPv6 and IEEE 802.15.4 Design Challenge

IoT solutions provide three network mechanisms: the uIP TCP/IP stack that provides IPv4 networking, the uIPv6 stack that provides IPv6 networking, and the Rime stack that is a set of custom lightweight networking protocols designed specifically for low-power wireless networks. The IPv6 stack was contributed by Cisco and was, at the time of release, the smallest IPv6 stack to receive the IPv6 Ready certification. The IPv6 stack also contains the RPL routing protocol for low-power loss IPv6 networks and the 6LoWPAN header compression and adaptation layer for IEEE 802.15.4 links. The Rime stack is an alternative network stack that is intended to be used when the overhead of the IPv4 or IPv6 stacks is prohibitive.

Mapping from the IPv6 network to the IEEE 802.15.4 network poses additional design challenges [14].

- *Addressing Management Mechanisms.* The management of addresses for devices that communicate across the two dissimilar domains of IPv6 and IEEE 802.15.4 is cumbersome.

- *Address Resolution.* IPv6 nodes are assigned 12-bit IP addresses in a hierarchical manner, through an arbitrary length network prefix. IEEE 802.15.4 devices may use either of IEEE 64-bit extended addresses or, after an association event, 16-bit addresses that are unique within a PAN.
- *Differing Device Designs.* IEEE 802.15.4 devices are intentionally constrained in form factor to reduce costs (allowing for large-scale network of many devices), reduce power consumption (allowing battery-powered devices), and allow flexibility of installation (e.g., small devices for body-worn networks). On the other hand, wired nodes in the IP domain are not constrained in this way; they can be larger and make use of mains power supplies.
- *Differing Focus on Parameter Optimization.* IPv6 nodes are geared toward attaining high speeds. Algorithms and protocols implemented at the higher layers such as TCP kernel of the TCP/IP are optimized to handle typical network problems such as congestion. In IEEE 802.15.4-compliant devices, energy conservation and code-size optimization remain at the top of the agenda.

2.1.5.7 IoT and Wi-Fi-Based Applications

When it comes to measuring technological performance, it is hard to resist the allure of a simple number. To compare digital cameras, buyers carefully compared megapixel count. For decades, clock speed – megahertz, then gigahertz – served as the universal shorthand for CPU performance. The higher the number the better it is, most people thought.

Of course, this was never true, and that became evident once single-core CPU performance plateaued. Manufacturers began touting their caching, their multicore processors, and their sophisticated bus architectures. The "megahertz myth" slowly lost its power. As a result, CPU developers began competing on the features of their entire platform.

Something similar is happening with Wi-Fi today. With the advent of the 802.11ac standard, Wi-Fi is encountering what we might call its "CPU moment." In the past, range and throughput were what mattered most in a Wi-Fi solution. But for next-generation Wi-Fi applications, those considerations are no longer sufficient.

That is not to say they are not important. They are critically important. But for emerging applications, like streaming wireless HDTV, rate and range do not mean anything without a third consideration: reliability.

In the past, Wi-Fi was a convenience. You needed it to check your email, work from home, update your iPhone. If you did not get a strong connection in one room, you could always move to another. It was annoying, but you could work around it.

Wireless HD and UHD video are different. Last year, Forrester estimated that nearly 115 million households in the United States owned at least one TV, and the number of households that watched online video on a TV set had increased by nearly 30%. A separate Forrester report estimates 66 million US households will access the Internet via game consoles, Blu-ray players or connected HDTVs by 2017. And increasingly, those connections will be wireless.

Delivering wireless HDTV today (and UHD coming up quickly) with wire-like quality requires a Wi-Fi signal that is fast, yes, but also reliable. According to a report from the OECD, by 2017, "households with two teenagers will have 25 Internet-connected devices." On the basis of current trends, most of these devices will be wireless, and all of them will contribute to interference.

For laptop users, an unreliable wireless connection might be an inconvenience. For TV viewers it is a potential deal-breaker. If they are having trouble with their picture, they cannot simply move their TV to a different room. They will not tolerate buffering. They will expect their TV to work the way it did when the signal came over cable. And they will expect to have this same experience with multiple TVs running at once.

As the end user's tolerance for error declines, the differences between Wi-Fi solutions become more important. While 802.11ac is sometimes described as "Wi-Fi for HDTV," not all 802.11ac solutions are equally capable of delivering an HD signal. These features, such as higher-order MIMO, are equally required to enable the performance in 802.11ac, just as they were in 802.11n. An access point with 4 × 4 MIMO could deliver twice the performance of a 3 × 3 MIMO device at equal transmit power, or the same performance using much less transmit power.

To penetrate walls and navigate around potential sources interference, precision is as important as power. To that end, digital beam-forming will be adopted on wider scale by 802.11ac products going forward. Digital beam-forming, as originally defined in 802.11n, had multiple optional modes, while 802.11ac limits the implementation options to only one, which will enable better interoperability. Digital beam-forming allows a transmitting device to aim its Wi-Fi signal at the receiver rather than broadcasting equally in all directions. This not only improves the Wi-Fi signal's range, but also its speed. But beam-forming is not a stock commodity. Different 802.11ac solutions from different manufacturers will have noticeably different beam-forming capabilities, determined by algorithms of different quality.

Different Wi-Fi solutions will prioritize HD video differently, as well. Some are effective at managing Wi-Fi streams to ensure that one slow device (often a mobile device) does not slow down the whole network. With others, a single misplaced access point could bring down the whole network. A device with multi-user 4 × 4 MIMO (MU-MIMO) will be even more effective at sending multiple video streams to multiple devices, akin to a wired connection.

Service providers are well aware that the days of "good enough" wireless performance are over. To meet the demands of 1080p HDTV, 4K UHD, and the increasing number of devices connecting to a home network (i.e., the IoT), ISPs are preparing to provide Wi-Fi technology that delivers a flawless high-speed signal for any purpose. Wi-Fi in the home is about to take a huge leap forward. You might not be able to measure it with a single number, but the difference will be unmistakable.

2.1.5.8 Inter-Cloud

A group of interrelated technologies is redefining how we live and work: cloud computing, big data, mobility, and the IoT. The cloud is at the epicenter of all this activity: big data migrates to the cloud to be sliced and diced; today's tablets, smartphones, and phablets rely on the cloud for services and entertainment ranging from social networking and microblogs to streaming video; and the hyper-connected world of smart grids, biosensors, and connected vehicles will rely on the cloud to collect data and then turn down thermostats, alert physicians, or avoid collisions.

As a result, cloud computing is growing exponentially. The industry is generally viewed as having three major segments: "infrastructure as a service," where computers and storage are available on an on-demand, pay-per-use rental basis, while residing in a cloud service provider's data center; "software as a service," which includes applications accessible over a network such as the Internet, for example, customer relationship management and billing but also voice and video communications and gaming; and a layer in between, referred to as "platform as a service," which enables developers to rapidly develop and deploy new applications.

A broad variety of companies offer cloud products and services, and just to list the largest ones would use up all of the space in this article. Suffice it to say that there are hundreds of cloud service providers of various sizes, and a rich ecosystem of additional companies that provide chips, networking gear, servers and storage, power distribution equipment, colocation and interconnection facilities, software stacks, management and monitoring tools, consulting services, and so forth.

This rich set of options offers plenty of choice to customers, whether they are consumers or businesses. And, as the cloud moves from early adopters to ubiquity, customers will leverage not just one cloud provider, but many, to solve needs across various business units and functions as well as one-off strategic initiatives.

However, a plethora of providers has its downsides: cost and risk associated with the management of complexity, search costs associated with provider selection, and transaction costs associated with everything from migration to operations.

And, while there will no doubt be some consolidation and organic growth, causing large providers to become even larger, there will also likely be a long tail of mid-sized and small providers that will compete on other factors, such as location, specialized expertise, or customer intimacy. By way of analogy, Walmart, Target, and Albertsons may have a substantial share of the retail market, but there is no shortage of "Mom and Pop" fruit stands. Marriott, Starwood, and Hilton may have a large portion of the global hotel market, but there are plenty of roadside inns and bed and breakfasts.

The IEEE is helping to ameliorate problems associated with incompatibility across multiple providers. The IEEE is a global organization with a charter to "foster technological innovation and excellence for the benefit of humanity." Among other things, the IEEE has developed hundreds of critical industry standards, including one that we are all familiar with, IEEE 802.11, better known as Wi-Fi. In 2012, the IEEE launched the Cloud Computing Initiative, chaired by Steve Diamond, to facilitate and coordinate cloud computing and big data activities across the breadth of the IEEE. An early focus of the Initiative was the formation of the IEEE P2302™ Inter-cloud standards working group, chartered to develop standards for cloud computing interoperability and federation.

In the same way that proprietary networks were made interoperable and thus easy to use by the Internet, the vision of this effort is for clouds, including proprietary clouds, to be made interoperable and easy to use via the Inter-cloud. The Inter-cloud does not supplant today's cloud providers; in fact, it should benefit current and emerging providers as well as business and consumer customers.

Analogies from the air travel industry and the Internet can help explain this. While there are hundreds, if not thousands, of airlines and millions of web sites around the world, they have agreed to jointly follow a number of protocols and procedures.

Airlines often need to transfer bags between carriers and the Internet needs to transfer data packets between ISPs. The Inter-cloud will need to reliably transfer applications and data between providers.

Airlines needed to agree on what codes to use for airports and the Internet needed a standard way of using URLs such as "http://insights.wired.com/." The Inter-cloud will need mechanisms to uniquely identify cloud service providers.

Travelers must present valid identification documents, payment methods, and boarding passes; the Internet and Inter-cloud both need mechanisms for identification, authentication, and payment.

Airlines want the ability to advertise their available seats and the prices for those seats on a variety of exchanges such as Travelocity and Expedia; web sites want to be found in search results; cloud providers will want to be able to advertise available resources on Inter-cloud exchanges.

The similarities are apparently endless, because abstractly, any service industry – airlines, Internet, retail, hotel, rental cars, telephony, and so on – faces similar challenges, and there are clear benefits to solving these kinds of challenges collaboratively while continuing to compete in other areas. An airline can compete on price, on-time arrival, service quality, and legroom. It does not need to compete based on a better scheme for naming airports.

And, in the same way that a shared mechanism for interline baggage transfer does not obviate the need for airlines, nor a standard mechanism for IP addressing put any web sites out of business, the Inter-cloud will facilitate continued growth of the industry, while increasing customer satisfaction.

The Inter-cloud will benefit service providers because they will be able to more easily partner to share each other's resources, advertise services to customers, participate in markets and exchanges, and help each other out during emergencies such as outages. It will benefit customers, because they will be able to more easily select from multiple providers, switch providers if one ceases service, offer improved user experiences thanks to a more dispersed footprint for highly interactive applications, and more easily build complex solutions from individual service components. Moreover, the Inter-cloud will take the onus for solving these problems away from users, helping to accelerate cloud adoption.

To help make the Inter-cloud standards a reality, the IEEE recently announced the Inter-cloud Test bed initiative to help evolve and validate the P2302 standard. Emerging standards need to be tested in the real world on real physical networks, servers, storage, and software. And interoperability standards require even more testing, as various combinations of components are evaluated for things like functionality, reliability, and performance. Almost two dozen companies and academic institutions have come together as founding members of this initiative and the test bed and related IEEE-led efforts are open to all who wish to contribute – not only service providers, but equipment vendors, subject matter experts, and end-customers.

The idea is that testing – both abstractly and against emerging customer scenarios – will identify needed improvements in the emerging standard, which can be addressed. These improvements will then be further tested. The Internet took decades to go from a vision of packet switching to where it is today. Between the IEEE, industry, and academia, one can hope that the vision of an Inter-cloud, conceived as early as 2009 by test bed initiative founder and chief architect David Bernstein, cloud computing initiative chair Steve Diamond, and their colleagues, is now getting the attention it deserves.

2.1.5.9 HMI – Touch, Feel, and Control

The Human Machine Interface (HMI) involves the interaction between users and equipment and provides control and operating status but it also defines the "look and feel" of the equipment and applications being used.

The HMI can incorporate many different approaches including switches, TFTs, GUIs, JPEG pictures, and so on. Many can be used together, but with users demanding more advanced control and monitoring options, the use of advanced interfaces is increasing even on the simplest of applications.

2.2 Application Requirements – Use Cases

This section will illustrate the application requirements by selecting some applications from the three market groupings: Individual, Industry, and Infrastructure.

The following are the identified applications:

- Health & Fitness (Lead Example)
- Video Surveillance
- Smart Home & Building
- Smart Energy
- Track & Monitor
- Smart Factory/Manufacturing
- Others: Smart Car, Smart Truck, Drone, Machine Vision, and Smart City.

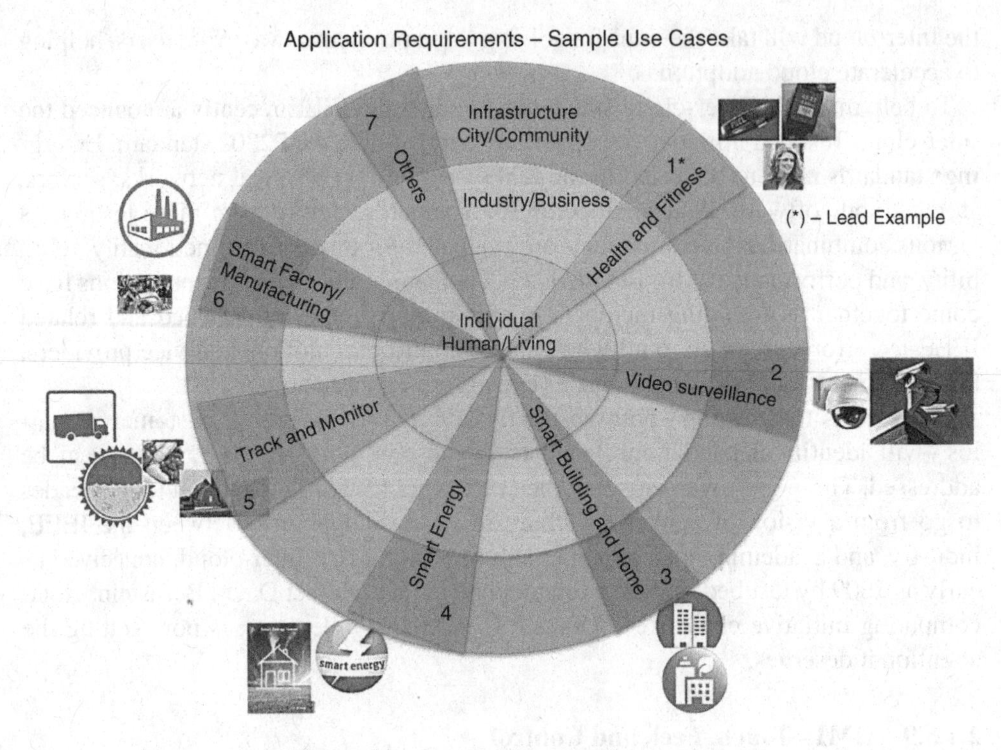

Figure 2.8 Application sample use cases

Sample applications across the three domains Individual, Industry, and Infrastructure are depicted in Figure 2.8.

2.3 Health and Fitness System for Individual/Industry/Infrastructure (LEAD EXAMPLE)

2.3.1 Landscape

Today, healthcare operates in silos across the three I's domains:

- *Individual.* Trying to be in the center striving to live healthy but is confronted with many challenges if he/she falls sick. Such an individual is frustrated because information is scattered, and is confronted with high cost of treatment and exuberant cost of surgery if the person is not covered by insurance.
- *Industry.* This is represented by here by physicians, clinics, labs, hospitals, and pharmaceuticals. Physicians are transiting from a paper-based system to an electronic system; typically, treat symptoms by prescribing medication, lab tests, and sometimes surgery. In general, physicians do not talk to those involved in alternate/functional medicine to understand the root cause and resort to natural means

to restore the body to a healthy condition and strive more toward wellness and prevention.

- *Infrastructure.* Here this is represented by the Government, which takes long to develop and administer policies and handle disputes between hospitals, physicians, insurance, and individuals. The Government takes time to release new drugs to the market and considers prescription drugs as mainstream, giving little attention in general to functional medicine and wellness.

These Health and Fitness functions and activities operate in silos. Figure 2.9 represents them, where some of the functions are centered on physicians, hospitals, or the Government. An Individual is confronted with a great challenge trying to live a balanced life with focus on Wellness/Prevention and absence of Sickness/Treatment. This is represented by the activities on the right-hand-side (RHS) of the circles versus those activities on the left-hand-side (LHS).

Call for collaborative IoT. As IoT begins to offer ways to have access to data resulting from sensing and other sources, tools gain insight from the accumulated set of

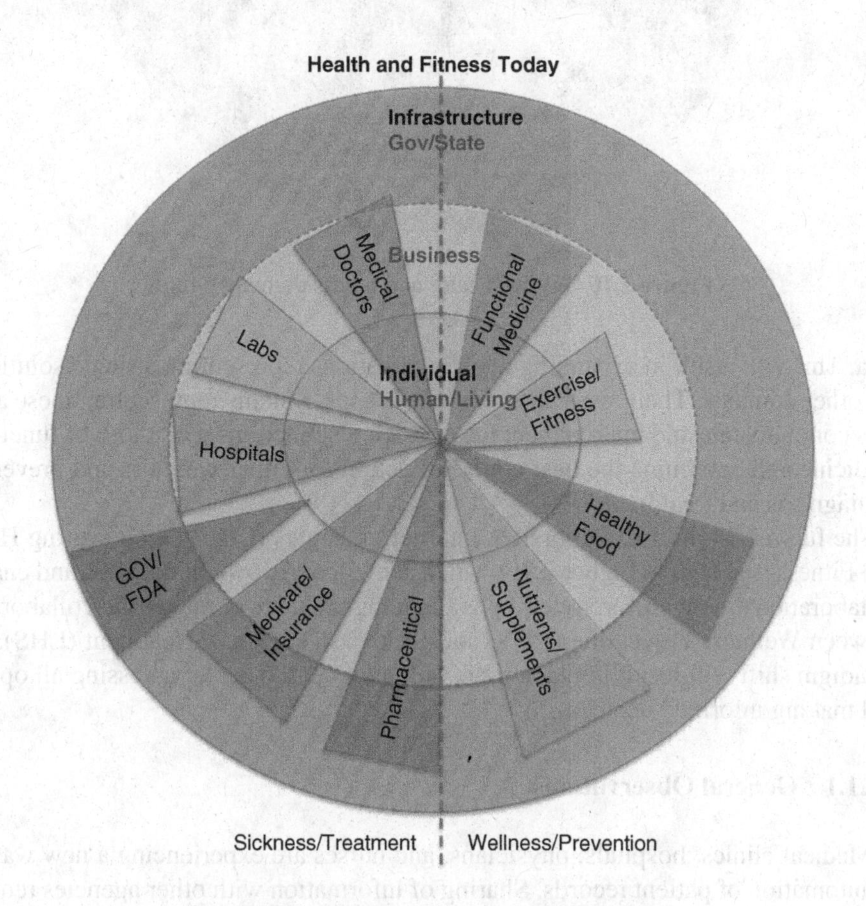

Figure 2.9 Health and fitness domains – today

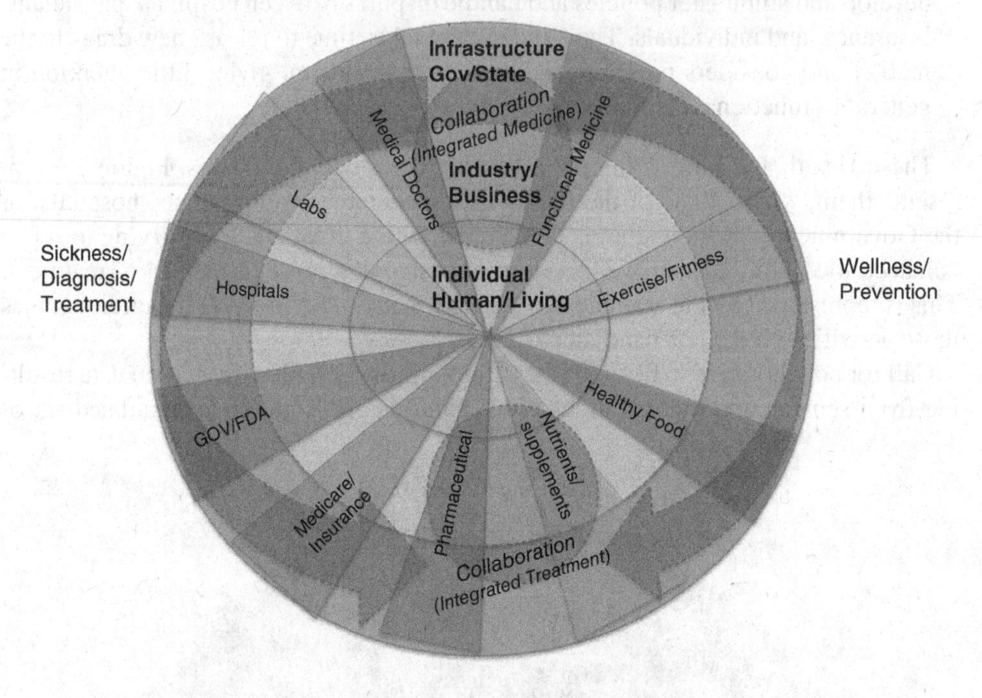

Figure 2.10 C-IoT health and fitness domains – future

data. This will result in expanding these applications representing a single solution to the other domains. Then, we start seeing these apps/systems representing these activities communicate and interconnect the physician's functions with those of functional medicine and determine the best course of action for either wellness and prevention or diagnosis and treatment.

The future vision of Collaborative Internet of Things (C-IoT) empowering Health and Fitness is shown in Figure 2.10; which shows breakdown of the silos and enables collaboration among different IoT apps spanning the three domains and collaborating between Wellness/Prevention (RHS) and Sickness/Diagnosis/Treatment (LHS). The paradigm shift will result in placing the Individual in the center assessing all options and making informed decisions.

2.3.1.1 General Observations

- Medical clinics, hospitals, physicians, and nurses are experiencing a new wave of automation of patient records. Sharing of information with other agencies requires compliance with the Privacy Act. IoT is helping the medical community to connect medical devices to the Internet and enable monitoring and tracking important information about patient.

- IoT is also playing an increasing role in the lives of individual where several initiatives have started allowing an individual to start monitoring and tracking health signs and fitness readouts about himself/herself that was never possible before.
- Such information will be of value not only to the individual but also to the physician under sickness condition. If surgery is required, then can the physician and patient view the readout from lab tests, along with tracking and recoding vital information about the patient? In addition, if a surgery is recommended, can pricing of surgery and insurance coverage be viewed by the patient for making a decision? Connecting these systems to establish a bigger picture is what IoT can enable over time.
- The Government plays a vital role in administering healthcare, connecting with physicians, insurance agencies, and administering Medicare/Medicaid services and policies and procedures.
- On the wellness/preventive side, IoT is also empowering individuals to monitor their burned calories, sleeping habits, and other factors. As technology advances and common platform development is in place, more IoT-based apps will be deployed connecting medical with wellness. Both will be aimed to ensure that the individual has a quality of life by being connected with all the parties concerned such as medical clinics, labs, pharmaceutical, hospitals, therapy, insurance providers, recreation centers, alternate medicine, and natural food stores.
- There are now 100 000 apps that are available in the Health & Fitness and Medical fields supported by platforms from Apple and Android. Both platforms are by far the leading mobile operating systems for mobile health apps today [15].
- The biggest group of mobile health apps is categorized as fitness apps. More than 30% of all apps that are listed in the Health & Fitness and Medical apps sections of Apple App Store, Google Play, BlackBerry Appworld, and WindowsPhone Store are fitness trackers or exercise guides.

2.3.2 Health & Fitness Sensing Requirements

2.3.2.1 Wearable Devices

Portable and wearable computing promise to introduce major shifts in how humans interact with computing devices and information, dramatically reducing the gap between immediate information and the person for whom that information is the most useful.

These small wearable devices come in the form of smart bracelets, smart watches, smart eye-glasses, smart T-shirts, smart shoes that are equipped with location sensors (RFID, NFCs (near-field communications), GPS) that track assets (kids, pets, elderly) as well as sensors for tracking health fitness biometrics (pulse, blood pressure, temperature, pedometer, etc.). Wearable wireless health devices also include accelerometers to warn of falls, and insulin pumps and glucose monitors for diabetics.

2.3.3 *Health & Fitness Gateway Requirements*

The adoption level shows an increasing trend and there will be more takers for these devices in the future. It is also evident that several software, service, and product companies are showing interest in connecting devices with a view to make their primary product or service more attainable.

Health fitness has many different dimensions of measurement, that it would be hard-pressed to imagine anyone spending an inordinate amount of time documenting these dimensions manually on their smartphones at frequent intervals. This is where the role of wearables comes in.

Wearable wireless fitness devices (bands) are another addition to the IoT. These connected bands take vital data from the body throughout the day and transmit wirelessly to user devices such as computers, smartphones, and tablets. These bands provide vital data to an individual and they are indeed a great tool to reduce medical expenses; health insurance companies too are taking interest in promoting them. Each of these devices can issue an alert to their physician by triggering a call over the cellular network.

Analysts expect the wearable computing device market to grow in the coming years. Juniper Research said in a report that the number of device shipments will increase from about 15 million devices in 2013 to 150 million devices by 2018.

2.3.4 *Health & Fitness Service Requirements*

C-IoT is gradually taking part in every facet of our lives. There is a high level of adoption of medical devices that are connected to the Internet and to each other. The recent emergence of a variety of wireless monitoring services is reaffirming this fact. Several connective devices have been already established in the healthcare industry.

Wearables need to go beyond simply measuring steps, heartbeats, and sleep cycles and attain the ability to measure the mood of individuals.

Today, a handful of companies produce wearable devices to detect brainwaves that infer how calm or attentive a person may be. With further growth in this area, sensors that are powerful enough to stream brainwave signals in real time will be developed. The form factors of these devices would also be compact enough so as to be inconspicuous.

2.3.4.1 Individual – Wearable Devices (e.g., Fitbit)

Consumers are now connecting their physical bodies to electronic devices such as Fitbit and other health monitors. Devices such as the Fitbit come with iOS and Android apps, include capabilities for social sharing, and track everything from sleeping habits, to the number of steps taken every day.

A fitness bracelet can also connect to another fitness device such as smart weight scale that monitors body weight, body fat, so the consumer's weight and fat are connected to the fitness-monitoring database for more accurate computing of calorie consumption, and so on.

In 2014, portable and wearable computing introduced major shifts in how humans interact with computing devices and information, dramatically reducing the gap between immformation and the person for whom that information is the most useful. Health and fitness buffs already wear monitors that record their heart rate and the distance they run, coupling that to a PC to analyze the results. Wearable wireless medical devices include accelerometers to warn of falls, and insulin pumps and glucose monitors for diabetics. Each of these devices can connect to a smartphone via Bluetooth, and can issue an alert to their physician by triggering a call over the cellular network.

Already since CES 2014, many products and reference designs for wearable devices have been released to the market. Examples include smart earbuds, smart headset, and a smart watch to a device – developed with Rest Devices for their Baby product line – that can be worn on an infant's onsite that monitors the baby's vitals and sends the data to a coffee mug, where it can be displayed.

Wearable Devices, Health Monitoring, and Your TV.

You are wearing a fitness/health device and watching television in your bedroom. The device is linked to your TV through shared logins or a simple linkage through your phone; so now, your TV knows about your body's activity levels including your sleep/wake patterns.

Figure 2.11 illustrates some examples of wearable devices.

Figure 2.11 Examples of wearable devices

Advertisers will make smarter TV buys and stop sending commercials to sleeping consumers. Taking this one step further, since your fitness device knows your activity levels, it could inform which commercials you see – for example, more active consumers might get the stepper and workout equipment commercials that, the author states, are wasted on him.

2.3.4.2 Wearable, Mobile, and TV

To build on the example above, looping your smartphone into this relationship will give advertisers an additional dataset. Your GPS and cell tower data tell us where you have been locally and if you have been out of town/on vacation. We know if you go to the gym, play tennis, or like to dine out.

On the basis of these specific data, digital (smartphone) and analog (TV) ads that are highly relevant to your behavior (and travels) can be sent to you.

2.3.4.3 Wearable Devices – Other Examples

- People are innovating on asthma inhalers (Propeller Health).
- Wearables (Pebble, Nike+) are opening up APIs for third-party developers.
- Hackers are combining IoT devices and APIs into amazing new use cases, for example, using a Fitbit activity tracker to pause movies on Netflix when you fall asleep or combining Fitbit with SmartThings to lock you out of your house until you have completed your morning run.

As largely a part of IoT, talking medical devices are continuing their journey effectively. This includes a device that reminds the person of taking a medication dose, checking blood pressure, taking a walk, or his/her cardio at a scheduled time. Examples of such groups that benefit from chatty clinical devices are those with regular age-related illnesses such as blood sugar and blood pressure and obesity patients. The way it works is that the app sends out notifications to the patients.

Wearables need to go beyond simply measuring steps, heartbeats, and sleep cycles and attain the ability to measure the mood of individuals.

Today, a handful of companies produce wearable devices to detect brainwaves that infer how calm or attentive a person may be. With further growth in this area, sensors that are powerful enough to stream brainwave signals in real time will be developed. The form factors of these devices would also be compact enough so as to be inconspicuous.

2.3.4.4 Performing Sentiment Analysis Using Big Data

With the human brain containing on average 86 billion neurons, there are potentially trillions of brainwave signals from these wearable devices that need to be analyzed.

The amount of time employees stay attentive at work can signify how engaged they are. A sustained series of spikes in brain activity could indicate stressful working conditions. Extreme brain focus at night, followed by lack of restful sleep may imply an organization with a workaholic culture.

Big Data algorithms will be able to correlate these data with several other bodily measurements such as sleep, physical activity, and heart rate to reflect the average sentiment within an organization in real time. Individuals could monitor their own sentiment privately within their smartphone app, and the general public could view anonymous aggregated data about an organization. Millenials, who already share data about their physical exercises through a combination of wearable devices and social media, will continue to do so with these highly advanced brain-sensing wearables.

2.3.5 Health & Fitness and Solution Considerations

Inter apps interface is required to support the following:

- Vital sign monitoring (blood pressure, heart rate, glucose, pedometer, etc.)
- Remote consultation and monitoring
- *Medical Consultation.* Medical reference apps provide information about drugs, diseases, and symptoms. The apps also give advice on how to take drugs or what to do in case of experiencing pain. They also show locations of pharmacies and medical centers/physicians.
- Diagnostics
- *Nutrition.* Nutrition apps help their users keep track of their diet, inform them about, for example, vitamins, calories, and fat content, as well as socio-economic aspects of food products (e.g., fair trade)
- Fitness and weight-loss trend analysis and alerts
- Reminders and alerts
- *Medical Condition Management.* Medical condition management apps represent the fifth largest group of mobile health apps. This group consists of all apps that track, display, and share the user's health parameters, medicament intake, feelings, behavior, or provide information on a specific health condition, for example, diabetes, obesity, heart failure.
- Personal health records (PHRs)
- Connectivity with providers (healthcare, insurance, family care giver)
- *Wellness.* Wellness apps summarize all kinds of relaxation solutions, yoga instructions, and beauty tips.
- Integration of local functionality and remote services
- Providing intelligence in gaining insights from big data via analytics tools and capabilities. Because of its powerful processing capabilities, storage capability, and diverse set of algorithms for cognitive computation, IBM Watson is emerging as an important system in analyzing vast amount of medical records.

2.3.6 Health & Fitness and System Considerations

- Integrated sensing, gateway, and services
- Context-aware, recognizing individual, device, location, and apps
- Personalized
- Diversity of user mobile/portable/fixed devices (smartphone, smart watch)
- Adaptive and responsive to change
- Anticipatory of subsequent steps and actions
- Secure at each layer sensing, gateway, and cloud.

2.3.7 Health & Fitness and Hospitals

Good hospital communications is essential for patient welfare and operational efficiency. With increasing numbers of events that need to be monitored, an increasing number of complex decisions need to be made to instantly notify increasing numbers of specialized medical responders via different wireless communications devices. These systems will become increasingly flexible to accommodate new technology as it emerges.

The wireless technology trend will follow the development of new types of sensor as they become capable of sensing the "chemical signatures" of an increasing number of treatable medical conditions.

It will also lead to an increase in automatic dispensing of medicine in response to what the sensor is reporting. For example, a small personal syringe pump could respond to the information coming from the sensor. This would enable smooth, precise dosing, without the unregulated "overshoot/undershoot" situation that may arise due to the delay between the test and the treatment.

2.4 Video Surveillance, Drone, and Machine Vision

2.4.1 Landscape

High-definition networked cameras, video surveillance software, and digitally based storage solutions are leading the industry in video surveillance. Right features and functionality would be required to meet basic surveillance needs for Individual home, Industry, or Infrastructure.

The next evolution is Proactive Video Surveillance Systems. These systems allow the user to intervene, manage, deter, stop, and apprehend perpetrators during an incident. Proactive systems are groundbreaking, providing new cost-effective tools that expand camera capability through video analytics and monitoring solutions.

Although video surveillance has increased in penetration at street intersections at different municipalities, demonstrating an irreversible trend for the future [16], we see

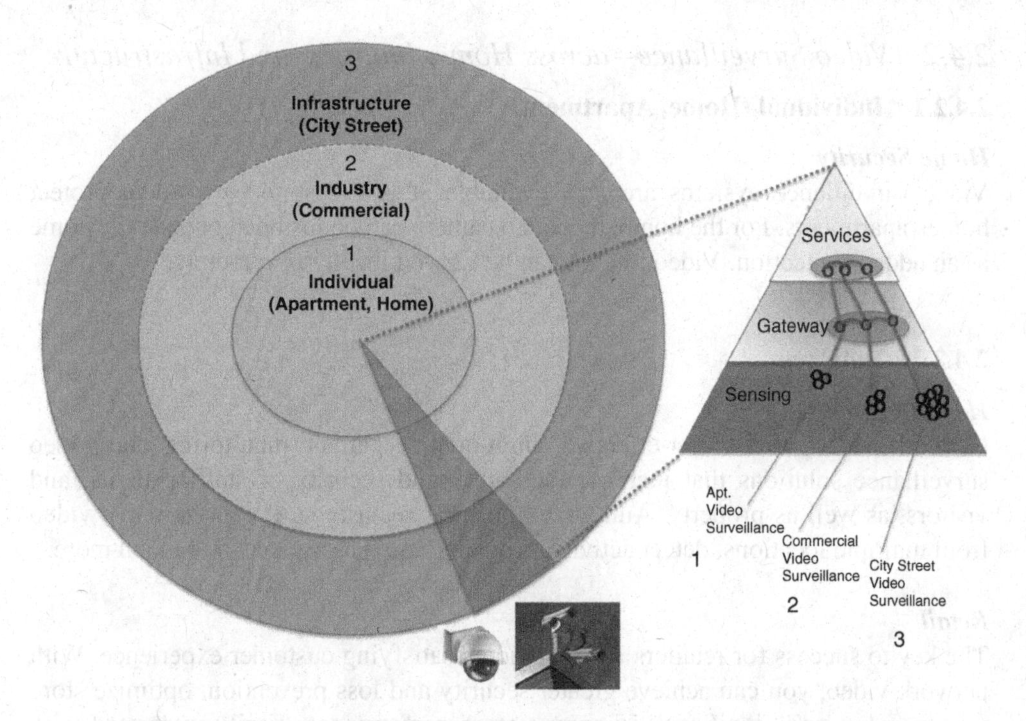

Figure 2.12 C-IoT video surveillance business App crossing three domains

video surveillance also penetrating home markets due to lower cost of cameras and overall system cost and this will continue to grow for various industry applications.

See Figure 2.12 for C-IoT video surveillance penetration of Individual, Industry, and Infrastructure domains.

2.4.1.1 Market, Drivers, Positioning

According to Transparency Market Research firm, Global Video Surveillance and Associated services (VSaas – Video Surveillance as a Service) market is $42.81 Billion by 2019 at a CAGR (compound annual growth rate) of 19.1% for 2013–2019. Regionally, it accounts for 35% share in North America, 31% share in Asia Pacific, and 44% in the Rest of the World. Asia Pacific is expected to be the fastest growing market with a market size of USD 17.12 billion in 2019.

Top target markets for video surveillance are commercial, industrial, institutional, residential, and infrastructural applications. Among the commercial applications, the office segment is observed to hold the highest share followed by infrastructure encapsulating highways, streets and bridges, transportation, communications, and stadiums.

2.4.2 *Video Surveillance – across Home, Industry, and Infrastructure*

2.4.2.1 Individual (Home, Apartment)

Home Security

Video surveillance systems are now available at low cost to be used to protect homes/apartments. For the home, the video camera can be installed outside the home as an added protection. Video images can be viewed locally or remotely.

2.4.2.2 Industry

Health Clinics

Network video offers cost-effective, high-quality patient monitoring and video surveillance solutions that increase the safety and security of staff, patients, and visitors, as well as property. Authorized hospital security staff can view live video from multiple locations, detect activity, provide remote assistance, and much more.

Retail

The key to success for retailers is to provide a satisfying customer experience. With network video, you can achieve greater security and loss prevention, optimize store management, and significantly improve store performance. Easily integrated with POS (point-of-sale) and EAS systems, an Axis network video solution enables remote and local monitoring at any time, from any place. You get rapid ROI (return on investment) as well as great interoperability; for instance, by combining people counting, integrated alarm functionality, and register monitoring.

Industrial

Network video is used in a multitude of industrial applications, such as remote monitoring of manufacturing lines and processes, performance enhancement of automated production systems, incident detection, and perimeter security. Network video can also support virtual meetings and improve remote technical support and maintenance.

Banking and Finance

Starting with existing CCTV equipment and infrastructure, you can create state-of-the-art network video surveillance systems that deliver exceptional image detail and powerful event management. Security staff can monitor multiple branches from a central or mobile location, and rapidly verify and respond to alarms.

2.4.2.3 Infrastructure

City Surveillance

Network video is an essential tool for fighting crime and protecting the public. In emergencies, network cameras can help police or firefighters to quickly focus their action. Advanced network cameras offer razor-sharp detail, motion detection, and tamper-resistance.

Operating over both wired and wireless networks, they are ideal and extremely cost-effective tools to promote the security that ensures safer cities.

Government

Network video helps protect all kinds of public buildings, from museums and offices to libraries and prisons. Supervising security at building access areas and remote monitoring points 24/7, Axis cameras increase security for staff and visitors. They also help prevent vandalism and provide visitor statistics.

Transportation

Network video gives you the means to improve safety, control flow, and enhance overall security at airports, railways, subways, and public transport hubs, and even on-board vehicles. Remote surveillance lets you monitor everything: check-in, platforms, gates, hangars, parking lots, and baggage systems, as well as vehicles in service. Network video traffic monitoring and management reduces congestion and improves traffic flow.

Education

From daycare centers to universities, Axis network video systems help deter vandalism and increase safety for staff and students. Using the existing IP infrastructure, no extra cabling is required. Features such as motion detection give security operators powerful tools to support action and avoid false alarms. Remote learning is another interesting application, for example, for students who are unable to attend lectures in person.

See Figure 2.13 for an example of C-IoT for video surveillance for a school campus.

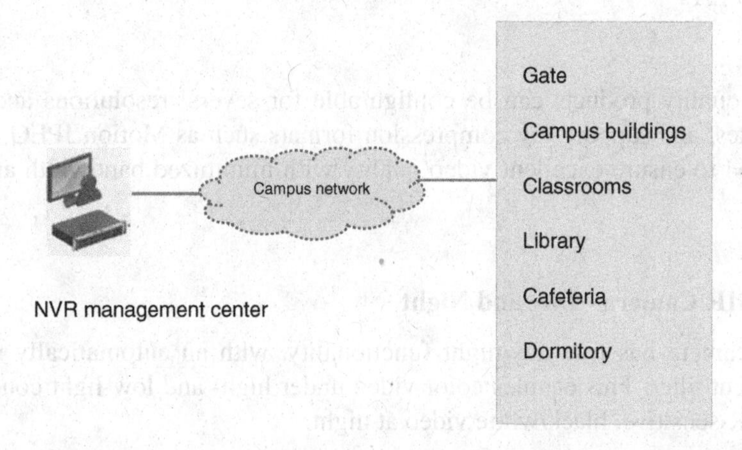

Figure 2.13 C-IoT and video surveillance for school campus

2.4.3 *Video Surveillance Sensing Requirements*

The IP camera is the information source provider of the video security system based on an IP network. It becomes more and more sophisticated with high-quality optics and digital video accuracy.

Mainly three functions are important in an IP camera:

- *Pixel resolution.* Pixel resolution such as high-resolution 720p is already available. Full HD is coming in the near future.
- Compression rate and H.264 AVC is also enough mature now and SVC will come soon.
- Performance head room for Video Content Analysis.

2.4.3.1 Lens Quality

Camera lenses are subjected to rigorous controls to assure the very highest quality in every respect.

2.4.3.2 Advanced Iris Control

P-Iris, a precise iris control system featured in some Axis cameras, further ensures optimal image quality in all lighting conditions.

2.4.3.3 CPU

A 32-bits CPU at 400 MHz is the minimum power to perform any image processing in addition to the IP encapsulation.

2.4.3.4 Video Coding/Encoding/Compression

- MPEG II/IV
- H.264

Video quality products can be configurable for several resolutions and different frame rates, and support key compression formats such as Motion JPEG, MPEG-4, and H.264 to ensure excellent video quality with minimized bandwidth and storage demands.

2.4.3.5 IR Camera – Day and Night

The IR camera has true day/night functionality, with an automatically removable infrared cut filter. This enables color video under high- and low-light conditions, as well as IR-sensitive, black/white video at night.

2.4.3.6 Power over Ethernet

The RJ-45 connector enables connection to the IP network. Built-in support for Power over Ethernet enables the camera to be powered via the network, consolidating power and reducing installation costs by eliminating the need for a power outlet.

2.4.3.7 Advanced Signal Processing

Advanced Signal Processing is used in designing ASICs for high-performance embedded applications. Areas of expertise include networking, open standards, Network protocols, and the most popular operating systems.

Different Wi-Fi solutions will prioritize HD video differently, as well. Some are effective at managing Wi-Fi streams to ensure that one slow device (often a mobile device) does not slow down the whole network. With others, a single misplaced access point could bring down the whole network. A device with MU-MIMO will be even more effective at sending multiple video streams to multiple devices, akin to a wired connection.

2.4.4 Video Surveillance Gateway Requirements

2.4.4.1 Cameras

- HD IP D1 Cameras (1080 × 720!) high quality images/videos
- Work in diverse operational environment (Weatherproof night vision, thermal)
- Hi resolution >3–5 Megapixel, wide viewing angle, long distance coverage, and zoom lens (2.8–12 mm!)
- Wall/pole/ceiling mount
- Some can be operated remotely for camera zoom, pan, and tilt
- Controller per cluster of cameras (8/16/32/64/128!)
- Potential upgrade to D4 with local storage and processing for facial recognition/license plate and capability to respond to real time query and real time alert of a match.
- Integrated audio signals into the video stream enabling two-way audio at remote locations
- Connectivity
- Networked-based camera connection for viewing and recording
- Potential advanced technologies for an Intelligent smart system that can detect unusual activity or compare images against database suspect photos
- Scalable system that expand to cover more locations and more cameras and enhance server capacity.

Figure 2.14 illustrates some types of IP image sensor cameras

Figure 2.14 Examples of image sensor IP cameras

2.4.4.2 Power Source and Cabling

- 120 or 240 V
- Backup battery
- Solar energy
- Power over Ethernet enabling connection to IP network.

2.4.4.3 Gateway

- Wireless
- Other options (HDMI output!)
- Concentrator/gateways/routers for public safety.

Determining capacity needs to take into consideration video demand on network bandwidth required, considering simultaneous streaming by cameras.

Facilitate audio/text message feedback to the target site via VoIP (voice-over-Internet Protocol).

There is a need for connectivity to a cloud server and capacity to store video images, and faster retrieval under flexible query.

The proposed networking equipment should include routers/gateways/concentrators, servers, and so on.

2.4.5 Video Surveillance Services

Fixed Position and PTZ (pan–tilt–zoom) video surveillance cameras provide live video pictures that are displayed on the GMS computer screen. This camera can capture a still picture when the gross weight is taken and this image can then be viewed on another work-station for comparison with a live video or saved to disk for future use.

Traffic Eyes are used to verify that the truck is correctly positioned on the scale. Two eyes and reflectors are used at the front and back of above-grade scales. Two additional long-distance eyes and sensors are used to monitor the sides of in-ground scales.

2.4.5.1 Requirements

- Video cameras are to stream videos to the cloud server and at the same time, video streaming can be viewed from a communication-monitoring center.
- If suspicious behavior is detected, the video monitor can send a public message via speakers mounted at key locations to deter the individual(s) and at the same time notify the first responder for action. Figure 2.13 illustrates an example of a communication-monitoring center and secure access.
- In some cases, camera can be installed in critical premises with a capability of a push button to connect directly to the central control for real-time viewing and intervention.
- Video clips can be downloaded about the incident covering, when, where, and how about the incident for further analysis and archived as a part of incident record system. The video is to be complemented by graphical location of the incident (GIS/GPS) and report from the first responder (Figure 2.15).

2.4.5.2 Local Storage

SD/SDHC memory cards enable recordings without having to use external equipment.

2.4.5.3 NVR – Network Video Recorder

Network of cameras redundant feed into 3 terabyte (SATA hard drive, Internal DVD!) – cloud server(s) and a shadow unit for hot-standby redundancy. Some

Figure 2.15 Communications monitoring center and secure access

cameras may have local storage but all will stream video into the NVR. The NVR has the capacity to hold feed up to 45 days before it is recycled (after being archived).

2.4.5.4 Remote Viewing

- Remotely view and manage your video surveillance system from anywhere through the Internet or smartphone/tablet in real time.
- Flexibility of viewing by sequence of camera, cluster of cameras, single camera, auto/on-demand viewing, and so on.
- Two-way audio at remote locations.
- Features/flexibility of viewing stored videos.

2.4.5.5 Management Platform

- Flexible surveillance management platform that can be integrated with other systems for a consolidated view of public safety.
- Application flexibilities for client custom features such as video analytics.

2.4.5.6 Visibility

Explore options to conceal visibility of cameras. Examples:

- Some camera systems make it more difficult for people under surveillance to determine if they are being watched, as it is usually impossible to figure out in which direction the camera is facing.
- Some cameras employ dummy lenses to conceal the surveillance target.
- Using a one-way transparent casing provides for the possibility of retaining the overt impression of surveillance – and hence a deterrent capacity – without having to place a camera in every housing or to reveal to the public (and offenders) the exact location under surveillance.

2.4.5.7 Service Plus

- The choice of camera locations should, ideally, result from a high-quality site assessment and analysis that not only incorporates a micro-level mapping of local targeted locations but also a potential scale of the system aiming at a bigger target.
- It is also valuable to conduct a number of site visits that examine the lines of sight for the cameras and identify any potential obstructions.
- If possible, construct a seasonal picture of the target site taking into consideration environmental seasonal changes and traffic.
- Identify hot spots and other areas that require coverage, location of 360° speakers, and wide-angle cameras.

- Ensure initial architecture and configuration is scalable to add more sites and regions.

2.4.5.8 IoT Hosting Services

Hosting solution would include installing the camera, connection to the Internet, maintaining the recording, and monitoring. A service provider will manage system maintenance as well as storage of recorded data. Typically, such services are ideal for small business at a single or multiple locations, small offices, retail stores, gas stations, and convenience stores. On the other hand, clients can now focus on the business and yet have remote access to the cameras from any location that provides an Internet connection, review recorded videos, and get event notifications.

2.4.6 Example: Red-Light Camera – Photo Enforcement Camera

- Red-light running is one of the major causes of collisions, deaths, and injuries at signalized intersections in the United States.
- Twenty percent of drivers do not obey intersection signals.
- Crashes caused by red-light running result in more than 800 fatalities and 165 000 injuries each year, according to the NHTSA.[1]
- The economic impact of red-light running on society is estimated to be $14 billion annually. Other motorists and pedestrians account for nearly half the deaths caused by red-light running crashes.
- Across the nation, communities are reducing the number of deaths and injuries from red-light offenses by 20–50%, simply by implementing red-light cameras.
- For 20 years, the technology of the photo enforceable camera has proven to be extremely accurate and reliable. Installed in over 500 communities.
- Does photo enforcement invade our rights of privacy "Big Brother" watching!
  - Photo enforcing technology is simply one tool available to the community to ensure that citizens are driving in a safe and responsible manner for the benefit of themselves and those around them.
  - When you choose to travel on public streets, you have a responsibility to operate in a safe and legal way.
- Does photo enforcement put police officers out of a job?
  - Photo enforcement helps ensure that limited resources are maximized.
  - Much like a radar gun used by an officer, photo enforcement is merely a tool that frees up some of the officers' limited time for enhanced safety and security in the community.
- Does my community really need red-light cameras?

[1] NHTSA (National Highway Traffic Safety Administration).

  - Insurance Institute of Highway Safety report, 27 January 2007 shows when red-light cameras are introduced, incidents of red-light running dropped from 198 incidents to 2.
- Will extending yellow signal timing be sufficient?
  - The use of adequate yellow signal timing reduces red-light running-related injuries and collisions.
  - Longer yellow timing used together with red-light cameras provides a more significant decrease in incidents of red-light running.
  - STAT: Results showed that yellow timing changes reduced red-light violations by an average 36%. The addition of red-light camera enforcement reduced red-light violations by an additional 96% beyond levels achieved by longer yellow signal timing alone.

See Figure 2.16 for an example of a red-light photo-enforced camera.

2.4.7 *Conclusion*

The market for network video products has grown tremendously in the past few years. The rapid deployment of network video indicates an irreversible shift from old, analog video technologies as network video advances with ever more effective, innovative, and easier to use products.

HDTV surveillance cameras are becoming the norm and more megapixel cameras are being introduced. There are cameras that can handle challenging lighting conditions such as low light, high contrast lighting, and total darkness, enabling improved surveillance capability. Processors in cameras and video encoders are not only faster

Figure 2.16 Example of red-light photo enforced

but also smarter. In addition, efficient video compression techniques as well as a new type of iris control, P-Iris, have been introduced.

Product choices are required to meet a variety of needs. There are smaller, more discreet – even covert – cameras, as well as thermal network cameras. There are needs for different fields of view, which are available from telephoto to 360° panorama. Axis' product development has also focused on easy and flexible installation. Outdoor cameras, for example, are weatherproofed right out of the box. Virtually most cameras and video encoders support Power over Ethernet, which simplifies installation. Many varifocal fixed cameras (box and dome) allow the focus and angle of view to be remotely set from a computer. Many fixed cameras also have the ability to stream vertically oriented views that maximize coverage of vertical areas such as aisles and hallways.

Managing cameras and video streams are being made easier. There is increased support for intelligent video functionalities. There are also video management solutions to suit every type of customer – whether it is a retail store with a few cameras or one involving hundreds of cameras at multiple sites. Products that support ONVIF can be easily integrated into systems that incorporate other ONVIF-conformant products from different manufacturers.

Greater network bandwidth is becoming more commonplace, and technologies have improved to make the transmission of data over wired and wireless networks safer and more robust.

Progress has also been made in storage solutions, especially for small systems. Today high-capacity network-attached storage (NAS) solutions are available that provide terabytes of storage at minimal costs and memory cards that enable weeks' worth of video to be stored in a camera or video encoder.

2.5 Smart Home and Building

2.5.1 Landscape

2.5.1.1 Smart Home

Smart Homes provides integrated, centralized control of two or more individual systems:

- Environment control
  - Smart thermostat (e.g., Nest)
  - Air moisture
  - Smoke alarm
  - Flood detector
  - Lighting system
  - Drapes
  - Sprinkler systems (outdoor)

- Smart Energy HVAC
  - Smart Meters
    Gas meter
    Electricity meter
    Water meter
    HVAC Control
  - Smart Appliance
    Refrigerators (consume 50% of home's energy budget)
    Stove
    Dishwasher
    Washer/dryer
  - Smart Plug
- Security/Safety (indoor, outdoor)
  - Video Surveillance (e.g., Dropcam)
  - Alarm System
    Motion Detectors Sensors
    Door Locks, key Fobs
    Window/Door/Garage control
- Health and Fitness
  - Wearable/portable devices (e.g., iWatch, fitbit)
  - Chronic Disease Management
  - Living Independently.

Networking/Multimedia

- Set Top Boxes and multimedia
- HD TV/Video Streaming
- Hi-Fi Systems
- VoIP
- Data Networking, Storage, Printing
- Remote Control.

See Figure 2.17 for C-IoT for a Smart Home.

2.5.1.2 Smart Building

- HVAC Control: heating (electric, gas), ventilation, air-conditioning
- Lighting control
- Smoke detector and sprinkler system
- Access control
- Data network, VOIP
- A/V system
- Wireless systems
- Facilities.

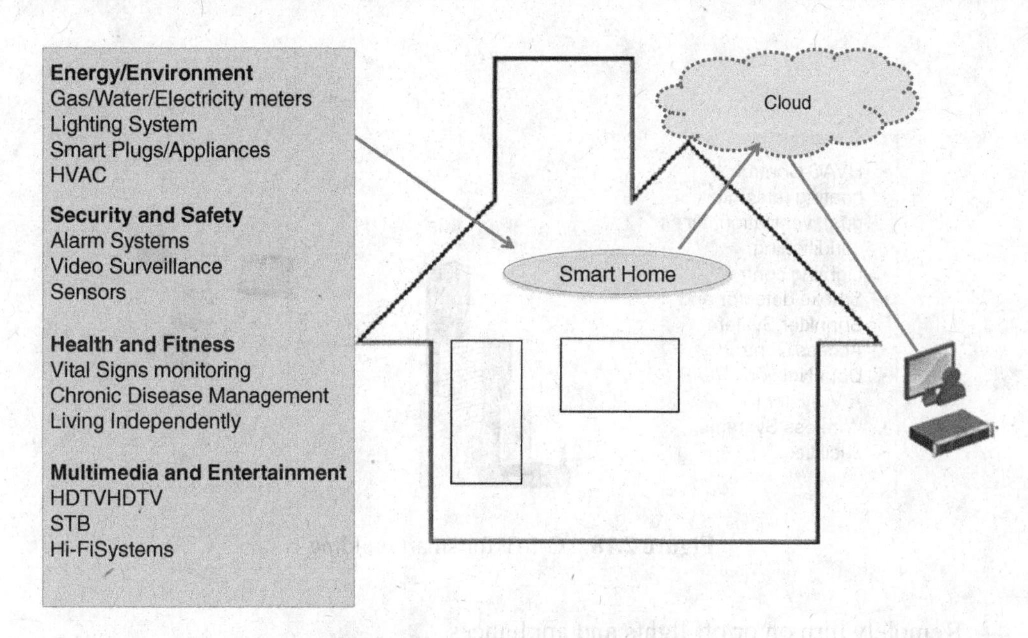

Figure 2.17 C-IoT for smart home

With an increasing demand for energy efficiency, safety, reliable connectivity, and precise control, industrial drives for factory automation systems are becoming more and more sophisticated, requiring cutting-edge technologies.

Devices: Smartphone, tablets, PC, wearable devices.

Devices may be connected through a wired or wireless network to allow control via a personal computer, and may allow remote access via the Internet (using a PC, smartphone, or tablet).

See Figure 2.18 for C-IoT for Smart Building.

2.5.2 Requirements

2.5.2.1 General Features

- Solution Architecture and platform, sensors, and other devices
- Integrated, wirelessly enabled platform that combines home security and automation capabilities
  - Monitor your home's energy usage and consumption
  - Remotely adjust your thermostat settings for heating and cooling
- Security monitoring centers 24/7
- Help lower your energy use by showing where/when electricity is being used
- Allow customers to customize a solution, based on individual needs, and the ability to manage and control their services from their location or from abroad
- Remotely lock or unlock doors

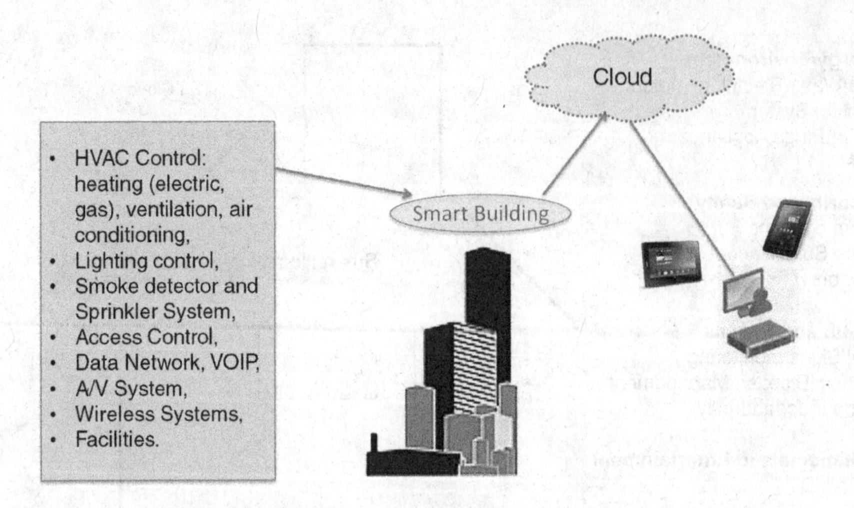

Figure 2.18 C-IoT for smart building

- Remotely turn on or off lights and appliances
- Set up to receive automatic event notifications
- The ability to add more features and services after the initial installation.

2.5.2.2 General Attributes

Many industrial network applications share attributes such as

- Real-time operation to detect state changes and take appropriate actions within an acceptable timeframe
- Deterministic operation to execute instructions in a predetermined order and at a predetermined time
- Reliable operation, often with $N+1$, $2N$, or $N+M$ redundancy, depending on the perceived cost of an outage
- Secure operation to ensure that unauthorized persons cannot accidentally or intentionally access or change data and manipulate control systems
- Safe operation to ensure that the system will not harm people or nearby equipment
- Ruggedized systems to operate in harsh environments such as
  - –40 to 120 °C temperatures at each chip on a board
  - Locations that are dirty, dusty, or surrounded by dangerous chemicals or emissions
  - Environments that contain high levels of electromagnetic radio emissions across a wide frequency spectrum
  - Remote systems, which are difficult to access for maintenance and repair. These systems require designs that minimize parts with higher failure rates, such as fans, to reduce time between system-level failures

– Systems that are operated by people who are not necessarily technology experts, who may not have access to sophisticated diagnostic equipment, and who often do not have time to read a long instruction manual or take a training course.

2.5.3 Smart Home & Building Sensing Requirements

2.5.3.1 Wireless Meters and Sensors

Affordable wireless sensors and meters can now be used to monitor automated building equipment and relay data to a centralized remote command center. Requirements call for

- Smart Meters for Electricity, Gas, Water, and Heat
- Grid Infrastructure
- RFID Transceiver + NFC smart Interface tag.

2.5.4 Smart Home & Building Gateway Requirements

An industrial gateway or industrial router is a ruggedized device that connects two or more networks, with signals directed only to ports where they are needed. The gateway can convert between standard Ethernet and industrial Ethernet protocols, between wireless and wired interfaces, or between Ethernet and fieldbus communications protocols. Processor performance can range from 200 to 1500+ MIPS. On-chip memory is often greater than 256 KB L2 cache. Factory automation equipment is ruggedly constructed for fanless operation in a harsh industrial environment.

2.5.4.1 Internet and Cloud Computing

The advent of the Internet and decreasing costs of data transmission now makes it financially feasible to transmit data from millions of building data points to the command center. The relatively affordable high-capacity computing power of the cloud allows for cost-efficient data analysis to an extent not possible in previous eras.

Gateway infrastructure securely aggregates data from a manageable number of meters and sends them to the utility servers.

2.5.4.2 Open Data Communication Protocols

Today, protocols such as 802.11ac, 802.11n Wi-Fi LAN, and 3G/4G WAN, integrated IPv4/IPv6 TCP/IP stack are required to support cross-platform data sharing. We have seen 802.11ac clients and APs (access points) come to market, but it will be some time before 802.11ac most new laptops and some smartphones can support it. Owing to the backward compatibility of 802.11ac with 802.11n, and because 802.11ac is limited to 5 GHz only, 802.11n will still be around for years to come. Already today, 802.11n is widely used to support rich services such as voice, video conferencing, and video streaming. More details about wireless technology and impact on applications trends can be found in [17].

Field bus protocols originally evolved to interconnect industrial drives, motors, actuators, and controllers. These numerous field buses include PROFIBUS, DeviceNet™, ControlNet™, CAN, InterBus, and Foundation Field Bus. Subsequently, manufacturers created higher-level networking protocols to interwork with the field bus protocols across Ethernet. These include PROFINET, EtherNet/IP™, Modbus® TCP, and SERCOS III. As a result, devices are needed to bridge between legacy and newer network protocols.

This brings opportunities for developers to innovate and challenge to meet Advanced SoC Architectural requirements for target market in the embedded space. For further details and examples, refer to [18, 19].

2.5.5 *Smart Home & Building Services*

The IoT-based solution will contribute toward reduction of operating costs, extension of product functionality, and enhancing user experience. The C-IoT-based solution for the connected home provides an integrated set of apps. These apps need to interoperate and work together harmoniously. For example, as you leave your work to go home and you press a single "coming home" button on your mobile device, in response, your house turns on the outdoor lights, brings up the indoor temperature from an energy-saving to a welcome-home temperature, closes drapes, and opens the garage door as soon you reach the premises and unlocks the door as you touch the handle.

2.5.5.1 Powerful Analytics Software

The best new-generation smart solutions provide numerous dashboards, algorithms, and other tools for interpreting building data, identifying anomalous data, pinpointing causes, and even addressing some issues remotely.

2.5.5.2 Remote Centralized Control

Secure Internet technologies can be used to protect data transmissions from hundreds of buildings in a company's portfolio to the central command center, staffed around the clock by facilities professionals.

2.5.5.3 Integrated Work-Order Management

Today's building management systems can be integrated with a work-order system to streamline communications with on-the-ground facilities staff when human attention is required.

In North America, companies such as Verizon, AT&T, Comcast, and Time Warner Cable provide home monitoring services for devices such as cameras, thermostat, appliances, and door/window control.

With connected refrigerators, lights, thermostats, and other sensors in the home, we can glean insights into the user's current environment. Is it dark? Is it warm or cold? Is the TV on or off? What are they watching? Who is home? This and more can be gained by connecting addressable advertising with connected homes.

Taking this one step further, sooner or later our fridge will know what we are low on, and send us ads or recommendations based on previous contents and consumption patterns. Our connected TVs will know what we watch and who is watching; this will allow marketers and networks to bring truly relevant content to the viewer's screen.

2.6 Smart Energy

2.6.1 Landscape

Most of our high-energy use today comes from heating/cooling, cooking, lighting, washing, and drying. These home appliances are beginning to become smart with connectivity features that allow them to be automated in order to reap benefits that smart metering and variable tariffs bring. The utility companies are beginning to better manage the energy demand and perform load balancing more efficiently.

On the Utility/Industry side, Data Aggregators/Concentrators form an important component in automatic meter reading (AMR). It creates the necessary network infrastructure by linking several utility meters (electricity, gas, water, heat) to the central utility server and captures and reports vital data. It also helps synchronize the time and date data of utility meters to a central utility server and enables secure data transfer of user authentication and encryption information. Communication to utility meters is comprised of an RF or wired (power line modem) connection, enabling data transfer to the central utility server via GPRS, Ethernet, and GSM, POTS, or UHF/VHF networks.

According to Pike Research, it is estimated that by 2020, there will be 963 million smart meter units and 63 million energy management users. This in turn will offer great opportunities for utilities to manage and control energy distribution to their customers; it also gives homeowners the opportunity to better manage their energy usage through smart energy management.

2.6.1.1 Individual: Smart iThermostat and Home

Smart homes will be able to self-manage things like energy consumption and climate control. Taking into account the homeowner's long-term energy goals as well as things like outside weather and city resources based on the smart grid, this future home will actively optimize its activity in real time.

2.6.1.2 Smarter Materials

Not only smarter, these homes of the future will be built with smarter materials. Currently in development, MIT's Mobile Experience Lab is building a new type of window that uses two layers of polymer-dispersed liquid crystal to let in light and heat when desirable and similarly, block it. Used in tandem with the structure's autonomous climate control, this facade will be able to shift to maximize comfort inside the home. Similarly, the prototype blends data from sensors both inside and outside of the home with historical weather information to determine the most comfortable combination of heat and air-conditioning while also minimizing carbon dioxide emissions.

- Cloud Storage and Computation
- Big Data Analytics
- Web and Mobile, remote operations
- Target Applications
  - Manage energy usage through smart energy management.

See Figure 2.19 illustrating C-IoT for Smart Energy.

2.6.2 Requirements

General requirements call to enable consumers to reduce electrical bills by deferring operation when energy costs are at a peak. In addition, energy patterns can be monitored so that the consumer can be given alerts if an appliance needs service, possibly avoiding breakdown and costly repairs.

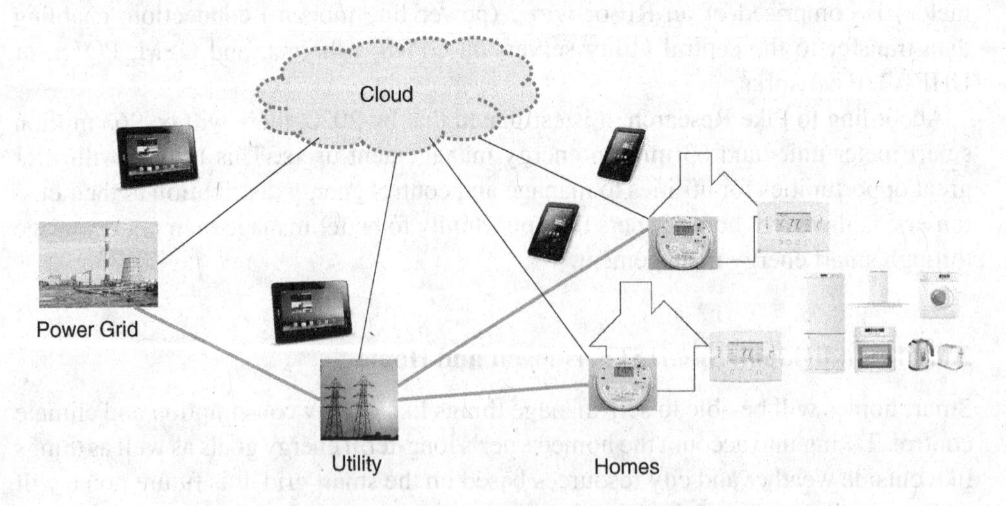

Figure 2.19 C-IoT for smart energy

2.6.3 Smart Energy and Sensing Requirements

- Climate control
  - Thermostats
  - Moisture detection
  - Thermostat, plugs, switches
- Appliance power controls (e.g., refrigerators)
- Security and alarm (motion and video, contact relays)
- Video surveillance cameras, motion detectors
- Garage door/Door locks control contacts.

2.6.4 Smart Energy and Gateway Requirements

- HAN
- WLAN – Wi-Fi
- Cellular WAN
- RFID
- ZigBee
- Siri.

2.6.5 Smart Energy – Services

IoT for Home Automation should focus on providing customers with a comprehensive home security and automation solution that offers the best possible customer experience and uses the most advanced mobile Internet technology on the market to make their lives easier and keep their families and property safer. The IT Services is to provide a unique suite of services, from start to finish, that will give homeowners control of their property and their possessions through an easy to navigate user interface.

- Established technology footprint, scale, and maturity (3B + Wi-Fi and 100 M HomePlug endpoints).
- Wi-Fi and HomePlug are based on the same IEEE networking model, which allows for seamless connectivity without translation.
- Combining the flexibility of Wi-Fi and ubiquity of the power line provides unsurpassed coverage throughout the HAN. When used in the HAN, the combination of Wi-Fi and HomePlug Green PHY provides an unsurpassed whole home coverage.
- High bandwidth (Mbps), IP-based connectivity enables state-of-the-art security protocols and quick software updates against cyber threats.

2.6.5.1 Smart Thermostat (e.g., iThermostat, Nest)

Smart Thermostat is a programmable thermostat, being offered by utility companies, that gives you control over the temperature in your home through any Web-enabled PC or smartphone for the purpose of helping customers save electricity and money.

The Smart Thermostat has evolved over time and a recent model includes a thermostat with a color touch screen display, an enhanced online portal with expanded features, and upgraded controls via the mobile app. It also automatically switches between heating and cooling as needed to keep indoor temperatures comfortable, a feature that is especially useful between seasons. The cost savings it offers just by using its preprogrammed settings complements the convenient controls.

The energy savings solution is available from utility companies to their customers, who receive the equipment and installation with no upfront charges. It leverages your high-speed Internet connection at home to connect your thermostat with the online- and smartphone-based controls.

2.6.5.2 The Smart Energy Dashboard

The utility company has expanded the features offering with Smart Energy Dashboard, an energy savings solution that offers a personalized view of your home's electricity consumption, giving you insight into how your habits at home affect your electricity usage and your bill.

The dashboard is available online to any customer for free by the utility company, and it now offers a new home comparison section that lets you compare the electricity usage of your home with similar homes in your area based on information you provide, including ZIP code, home type, age of home, square footage, and heating type. It also offers a new usage breakdown section including a pie chart that shows your approximate electricity usage and cost for heating, cooling, water heating, lighting, electronics, appliances, refrigeration, and other equipment.

The dashboard shows you how much electricity you are using over time and how the outside temperatures may be affecting your usage. It shows your electricity usage and cost by month, day, and even hour if you have a smart meter. It also offers you a forecast of your estimated usage and costs at any point of a billing cycle.

2.6.6 The Smart Energy App

Utility companies are also offering customers mobile access to their accounts through a free app on smartphones or tables. The app now allows you to control your home temperature settings.

It also provides a number of features available through the dashboard in addition to giving you the opportunity to check and pay your bill.

There are several other types of smart thermostat such as Nest (Google) and Lyric (Honeywell).

Nest (purchased by Google for $3.2 billion) is a company that makes a Learning Thermostat that saves energy with control over Wi-Fi from your mobile device or laptop. Nest also remembers what temperatures you like and turns itself down when you are away. Nest creates a personalized schedule based on the temperature changes you have made. All you have to do is get comfortable. It is so simple that 95% of Nests have schedules. Other features offered by Nest include:

- Automatically turns itself to an energy-saving temperature when you are away, so you do not waste energy heating or cooling an empty house.
- Connect your Nest thermostat to Wi-Fi, download the free Nest Mobile app, and change the temperature from anywhere.
- Provides Energy History and Report to understand how your energy use changes month-to-month and how you can save more.
- Shows you when you are choosing a temperature that will help you save. Look for the Leaf every time you turn the ring.

Recently, Nest has opened its platform to other devices and developers to develop IoT applications that enable interoperability between devices in the home using Google's cloud.

2.6.7 Smart Energy and Network Security

NIST SP 800 82 (NERC and IEEE) provides guidance for securing ICSs, including supervisory control and data acquisition (SCADA) systems, distributed control systems (DCSs), and other systems performing control functions. The guideline provides overview of ICS and typical system topologies, identifies typical threats and vulnerabilities to these systems, and provides recommended security countermeasures to mitigate the associated risks.

Major ICS Security Objectives

- Protecting individual ICS components from exploitation
  - This includes deploying security patches in as expeditious a manner as possible, after testing them under field conditions; disabling all unused ports and services; restricting ICS user privileges to only those that are required for each person's role; tracking and monitoring audit trails; and using security controls such as antivirus software and file integrity checking software where technically feasible to prevent, deter, detect, and mitigate malware.
- Maintaining functionality during adverse conditions
  - This involves designing the ICS so that each critical component has a redundant counterpart. Additionally, if a component fails, it should fail in a manner that does not generate unnecessary traffic on the ICS or other networks, or does not cause another problem elsewhere, such as a cascading event.

2.6.7.1 NIST SP 800-82, Rev 2

NIST SP 800-82, Rev 2 is a major update, which includes

- Updates to ICS threats and vulnerabilities

- Updates to ICS risk management, recommended practices, and architectures
- Updates to current activities in ICS security
- Updates to security capabilities and technologies for ICS
- Additional alignment with other ICS security standards and guidelines.

ICS overlay for NIST SP 800-53, Rev 4 security controls will provide tailored security control baselines for Low, Moderate, and high impact ICS.

2.7 Track and Monitor

Tag (and GPS) – Smart Tracking and Monitoring of Assets.

2.7.1 Landscape

2.7.1.1 RFID – Definition/Standards

In the IoT paradigm, many of the objects that surround us will be on the network in one form or another. RFID and sensor network technologies will rise to meet this new challenge, in which information and communication systems are invisibly embedded in the environment around us.

RFID refers to a wireless system comprised of two components: tags and readers.

2.7.2 Track and Monitor – Sensing Requirements

The reader is a device that has one or more antennas that emit radio waves and receive signals back from the RFID tag. Tags, which use radio waves to communicate their identity and other information to nearby readers, can be passive or active, and are attached to the object to be identified or tracked. Passive RFID tags are powered by the reader and do not have a battery. Active RFID tags are powered by batteries.

A typical operation consists of an RFID reader that transmits an encoded radio signal to interrogate the tag. The RFID tag receives the message and then responds with its identification and other information.

RFID tags can store a range of information from one serial number, a license plate, or product-related information such as stock number, lot or batch number, and production date to several pages of data. Readers can be mobile so that they can be carried by hand, or they can be mounted on a post or overhead. Reader systems can also be built into the architecture of a cabinet, room, or building.

RFID tags contain at least two parts: an integrated circuit for storing and processing information, modulating and demodulating an RF signal, collecting DC power from the incident reader signal, and other specialized functions; and an antenna for receiving and transmitting the signal. The tag information is stored in a nonvolatile memory. The RFID tag includes either a chip-wired logic or a programmed or programmable data processor for processing the transmission and sensor data, respectively.

Fixed readers are set up to create a specific interrogation zone, which can be tightly controlled. This allows a highly defined reading area for when tags go in and out of the interrogation zone. Mobile readers may be handheld or mounted on carts or vehicles.

Recently, decreased cost of equipment and tags, increased performance to a reliability of 99.9% and a stable international standard around UHF passive RFID have led to a significant increase in RFID usage.

Overview/Market/Drivers of Growth/Positioning/Relation to IoT (Reference IoT Model).

It could be argued that the industry got a little ahead of itself in the early 2000s when RFID was expected to transform supply chains overnight. The concept of the "IoT" was still years away, but the appeal of digitizing physical assets and achieving absolute visibility led to some grand theories. However, although much of the early hype has faded, RFID continues on a trajectory toward widespread adoption and impact.

According to RFID market research from IDTechEx, RFID market is on a very healthy evolutionary track, growing from $7.88 billion in 2013 to $9.2 billion in 2014. IDTechEx expects the RFID market will reach $30.2 billion in 2024 [20].

The driver to this growth is not the wholesale replacement of bar codes with RFID tags, nor is it the deep pockets of sophisticated retail, healthcare, or governmental entities. It is the mind-set that a mere handful of tags and a single reader can provide a meaningful ROI for a very small slice of a given process.

2.7.3 *Track and Monitor – Services*

When tags are in place because one part of the supply chain sees the value, the technology becomes more appealing for others to collect data because the required investment is much less. There is an opportunity to leverage an investment that someone else has already paid for, opening the door for more applications that contribute to streamline the operational efficiency and provide large amount of data to drive greater insights for establishing next course of action or driving the next phase of a roadmap.

RFID technology helps in automatic identification of anything they are attached to, acting as an electronic barcode. The passive RFID tags are not battery powered and they use the power of the reader's interrogation signal to communicate the ID to the RFID reader. This has resulted in many applications particularly in retail and supply chain management. The applications can be found in transportation (replacement of tickets, registration stickers) and access control applications as well. The passive tags are currently being used in many bank cards and road toll tags, which is among the first global deployments.

Active RFID readers have their own battery supply and can instantiate the communication. Of the several applications, the main application of active RFID tags is in port containers for monitoring cargo.

2.7.4 *Track and Monitor – Solution Considerations*

RFID is not bar codes. Bar codes will remain the standard for capturing data at specific points in a process for years to come. It is the places between those points where RFID might prove valuable. RFID will tell you what happened for an object being tracked from point A to point B. Now you know both where an object is and what it is doing. From there, you can start to think about what it could be doing. Unlike a bar code, an RFID tag constitutes a unique identifier for the tagged item – it is not those kids, elderly, trucks, trains, but it is those specific uniquely identified kids (by name) in schools, elderly (by name) at home, trucks (by ID) at warehouse, or trains (by unique ID) at station. Whether between two points in a facility or between two cities, RFID can prevent objects (products and assets) from getting lost along the way. In the process, it can create visibility from Point A to Point B or from a warehouse – dock door to doorstep of a retail store.

Uniqueness is often important in e-commerce applications where value-added information results in better comprehension that can be targeted to achieve greater efficiency.

Because RFID is for relatively short ranges, from between sub-meter to a few meters, it is complimentary to GPS technology for both asset and product tracking. For instance, a container might be fitted with GPS and an RFID reader and as an RFID-tagged case is put on the container the two might be associated in software. When paired with RFID technology, sensors and data loggers can monitor conditions like temperature and impacts. These are widely used in the food traceability industry, mainly to meet insurance requirements.

Active RFID is nearly the same as the lower-end WSN nodes with limited processing capability and storage. The scientific challenges that must be overcome in order to realize the enormous potential of WSNs are substantial and multidisciplinary in nature.

2.7.5 *Track and Monitor Examples*

2.7.5.1 Implanted RFID Chip to Manage Critical Healthcare Issues

RFID tag implanted under the skin of a person empowers the individual to know and manage his/her healthcare issues. The tag could alert physicians of the implants during surgery and relay necessary information for individuals with life-threatening diseases, and could be particularly useful during a medical crisis. It also could be used to keep track of other implantable devices, such as a pacemaker, a patient might have. The tag does not contain any medical records, but its 16-digit number could be linked to a database of patient medical information. When the tag is scanned, the number could be quickly cross-referenced to reveal specific medical data about the patient. The 134.2-KHz RFID tag could save lives and possibly limit injuries from errors in medical treatments.

2.7.5.2 Advancement beyond RFID Tags to Track Elderly at Home

A lot of elderly people live alone. So if they get into trouble (e.g., fall down unconscious or have an injury that prevents them from moving) there is no one aware that they need help. If the environment in their households can be automatically monitored for signs of sudden onset of an acute health problem then helpers can be sent to check on them, video cameras could be activated to see if they are all right, or they could get a phone call.

An RFID reader-tag system can be used to track elderly. Researchers are adapting RFID and sensor technologies to automatically identify and monitor human activity to be able to determine if an individual's normal routine is being maintained so that timely assistance can be provided if it is needed. Home medical monitoring can and will go far beyond what RFID tags can accomplish. Imagine an electronic monitoring system built into your bed that monitors the gases in your breath, tracks your breathing and pulse, and studies your movements in bed. It could detect problems like sleep apnea or nervous disorders and diagnose a chronic illness in its early stages.

2.7.5.3 Safety of Our Kids in Schools and Buses

RFID-enabled badge Readers are being installed on schoolbus doors and in classrooms. When a child is in range, the reader transmits data to a GPS system. The system is only being used for notification purposes, giving parents and schools the ability to log on and view students' attendance records.

The schools across the country are adopting a variety of different tools to monitor students both in school and outside school. Among these tools are RFID tags embedded in school ID cards, GPS tracking software in computers, and even CCTV video camera systems. According to school authorities, these tools are being adopted not to simply increase security, but to prevent truancy, cut down on theft, and even improve students' eating habits.

2.8 Smart Factory

2.8.1 Factory Automation – Robot

Robots fascinate us. Their ability to move and act autonomously is visually and intellectually seductive. We write about them, put them in movies, and watch them elevate menial tasks like turning a doorknob into an act of technological genius.

For years, they have been employed by industrial manufacturers, but until recently, never quite considered seriously by architects. Sure, some architects might have let their imaginations wander but not many thought to actually make architecture with robots. Now, in our age of digitalization, virtualization, and automation, the relationship between architects and robots seems to be blooming

- By the end of 2007, there were around 1 million industrial robots in use, worldwide. About 60% of these are articulated robots, and about 22% are gantry robots.
- The global market for industrial robots was worth approximately $6.2 billion in 2008, excluding software, peripherals, and systems engineering.
- The automotive industry is the largest user of industrial robots.
- Although the global economic crisis has severely impacted sales of industrial robots – especially in the key automotive sector – the need to automate industrial processes to improve efficiency and safety of the workplace remains constant, and is expected to lead to renewed growth between 2010 and 2012.

2.8.2 *Industrial*

Industrial Robotics is a branch of engineering that combines electronics, control systems, mechatronics, artificial intelligence, computer science, and bioengineering. A robot is a device, which includes sensors, actuators and a control system. Robots are generally classified by their purpose:

- A factory robot or an industrial robot performs jobs such as cutting, welding, and gluing;
- A service robot or a mobile robot adds to its primary tasks also movement within its working environment.

See Figure 2.20 for an example of a robot in smart factory.

Embedded Power provides a complete suite of power conversion solutions to meet the needs of robotics applications.

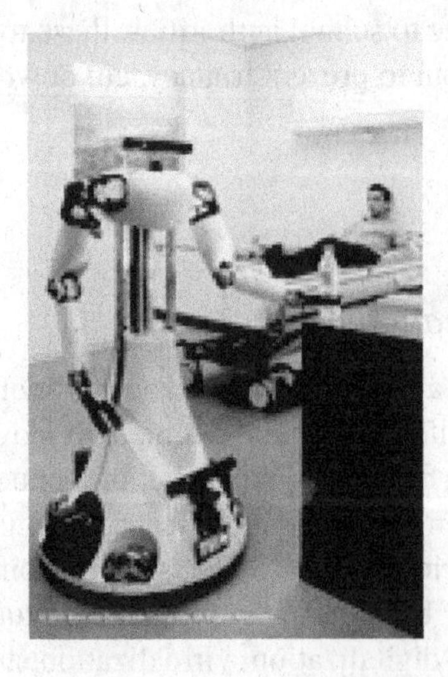

Figure 2.20 Future C-IoT robot as a caregiver

- Automated assembly
- Automated manufacturing
- Automated packaging
- Warehouse management
- Goods handling
- Goods transport
- Pick-and-place systems
- Portal robots.

The following are the key enabling product requirements:

- Integrated ADC and FPU and analog
- Standard Low-side, High-side, and Bridge Smart Power Devices for driving solenoids, DC motors, and stepper motors
- Dedicated ICs for actuator driving, charging, and power management
- One of the industry's broadest ranges of Power MOSFETs and IGBTs (Insulated-Gate Bipolar Transistors)
- PC Controller technology
- Real-time Operating Systems based on multiprocessor technology
- Safety Technology – 100 Mb, multicore processor technology
- Open interface

2.8.2.1 Robot Arm – Multitude of Sensors

A robotic arm assembly may have movement in several axes. A worm-driven shoulder joint could move the arm up and down in conjunction with an electric-driven hydraulic actuator. An independent elbow joint adds another axis of control, and a wrist joint adds a directional and rotational set of axis.

While each of these can be an independent set of control loops using their own stress and strain sensors, they are not isolated functions. The movement of an arm up, for example, can induce overload stresses on a wrist joint if too much weight is loading down the arm and the wrist is in the wrong position. As a result, all sensor data can be important, even for sometimes seemingly unrelated tasks. This is the fusion of sensory data where the big picture is looked at rather than an individual control loop.

Example: Multi-fingered robot hand for industrial robotics application.

A wireless teleoperated robotic hand system is intended for providing solutions to industrial problems like

- Robot reprogramming,
- Industrial automation, and
- Safety of the workers working in hostile environments.

The robotic hand system works in the master slave configuration where Bluetooth is used as the communication channel for the teleoperation.

The master is a glove, embedded with sensors to detect the movement of every joint present in the hand, which a human operator can wear.

This joint movement is transferred to the slave robotic hand, which will mimic the movement of human operator. The robotic hand is a multi-fingered dexterous and anthropomorphic hand.

All the fingers are capable of performing flexion, extension, abduction, adduction, and hence circumduction. A new combination of pneumatic muscles and springs has been used for the actuation purpose. As a result, this combination reduces the size of the robotic hand by decreasing the number of pneumatic muscles used. The pneumatic muscles are controlled by the opening and closing of solenoid valves.

2.8.3 Service Robot

According to the International Federation of Robotics (IFR), a service robot is a robot that performs useful tasks for humans or equipment excluding industrial automation application. Note: The classification of a robot into industrial robot or service robot is done according to its intended application.

- A personal service robot or a service robot for personal use is a service robot used for a noncommercial task, usually by lay persons. Examples are domestic servant robot, automated wheelchair, personal mobility assist robot, and pet-exercising robot.
- A professional service robot or a service robot for professional use is a service robot used for a commercial task, usually operated by a properly trained operator. Examples are cleaning robot for public places, delivery robot in offices or hospitals, firefighting robot, rehabilitation robot, and surgery robot in hospitals. In this context, an operator is a person designated to start, monitor, and stop the intended operation of a robot or a robot system.

A degree of autonomy is required for service robots ranging from partial autonomy (including human–robot interaction) to full autonomy (without active human–robot intervention). Therefore, in addition to fully autonomous systems, service robot statistics include systems that may also be based on some degree of human–robot interaction or even full teleoperation. In this context, human–robot interaction means information and action exchanges between human and robot to perform a task by means of a user interface.

In some cases, service robots consist of a mobile platform on which one or several arms are attached and controlled in the same mode as the arms of industrial robots. Furthermore, contrary to their industrial counterparts, service robots do not have to be fully automatic or autonomous. In many cases, these machines may even assist a human user or be teleoperated [21].

Figure 2.21 Robot in smart factories

2.8.3.1 Caregiver Robot

Robots may become gentler service/caregivers in the next 10 years. See Figure 2.21.

Lifting and transferring frail patients may be easier for robots than for human caregivers, but robots' strong arms typically lack sensitivity.

Japanese researchers are improving the functionality of the RIBA II (Robot for Interactive Body Assistance), lining its arms and chest with sensors so that the robot lifts and places patients more gently.

2.9 Others (Smart Car, Smart Truck, Drone, Machine Vision, and Smart City)

2.9.1 Smart Car

2.9.1.1 Definition

Next-generation telematics and infotainment for passenger cars are not just around the corner, they are here.

Embedded wireless connectivity enables a broad range of functionality that is transforming the automobile experience. From basic safety and security services, such as automatic crash detection and notification, to next-generation safety services enabled by vehicle-to-vehicle (V2V) and vehicle-to-roadway infrastructure communications, to enhanced remote diagnostics and maintenance, to 4G LTE-powered advanced 3D navigation and connected infotainment services, mobile technologies are at the forefront of driving innovation in the next generation of connected cars.

This automotive heritage combined with unparalleled expertise in mobile technologies allows it to play a unique role in enabling automakers to offer advanced systems that empower drivers and passengers with the capabilities they have come to expect from their connected consumer devices and more, while truly revolutionizing the driving experience.

To address the rapidly expanding automotive opportunity, Automotive Solutions, comprising automotive-grade Snapdragon processors, Gobi 3G/4G LTE multimode modems, and low-power Wi-Fi and Bluetooth solutions, provide unprecedented, integrated connectivity options for the automotive ecosystem to create new, breakthrough connected systems and services.

Landscape

- Standard car/efficiency and IoT empowered for efficiency, safety
- Electric car
- Driverless car
- V2V
- Vehicle-Infrastructure

See Figure 2.22 about C-IoT for smart cars.

2.9.1.2 Market, Drivers, Positioning

Vehicle and C-IoT

According to Semicast Research, the "*revenues* for original equipment (OE) *automotive semiconductors* grew by *12% to USD $25.5 billion in 2012*, while the *total semiconductor industry* is *declined by almost 3% to USD $292 billion*." This represents a high growth in the semiconductor automotive industry compared to the negative growth of the total semiconductor market [22].

The automotive world is changing rapidly. The unprecedented growth is driven by advancement of technology in the automotive technology empowered by IoT-enabled sensors, Robot-assisted manufacturing, and larger adoption.

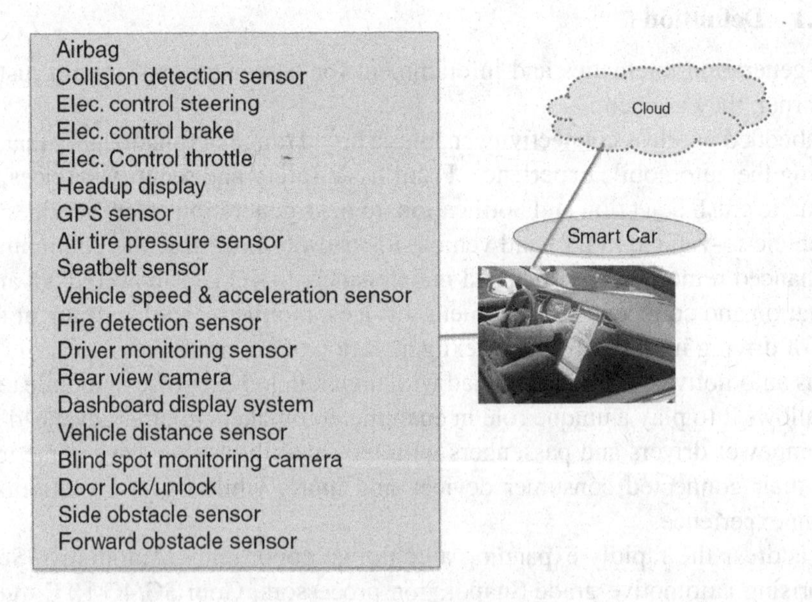

Figure 2.22 C-IoT for smart cars

Advancement of technology in alternate energy has led to the introduction of electric and hybrid vehicles. Six-month sales data suggest that in the first half of 2013, all-electric vehicle sales in the United States unexpectedly overtook sales of plug-in hybrid electric vehicles (EVs that also have a combustion engine) for the first time.

In addition, most car manufacturers are developing "autonomous cars," that is, vehicles that still need a driver to take over the steering wheel or acceleration/braking pedal functions in case of unanticipated events. But Google has made a leap with a pure "driverless car" – a car without human intervention, steering wheel, and acceleration/braking pedals. This leap will boost revenue growth for semiconductors over the coming years. It is important to note that unlike today's autonomous cars; Google's driverless car totally depends on its sensors, semiconductor ICs, and algorithms running inside several electronic control units (ECUs). Semiconductor technology is already available for implementing driverless cars, but human acceptance is a key challenge.

CAR – Head-Up Display

Aerospace has a continuous influence into the design – aerodynamics of cars and now in terms of instrumentation and guidance system. The latest is head-up display (HUD) technology, which was originally developed for fighter jets. Projecting information directly into the driver's line of sight allows people to process it up to 50% faster – due to shorter eye movement – and keep their attention focused on the road ahead. Head-up display technology is currently available on several high-end vehicles and is starting to show up in other segments. In fact, more than 35 vehicle models currently available in the United States have standard or optional HUDs. According to IHS Automotive, 9% of all new automobiles in 2020 will be equipped with HUD technology versus 2% in 2012. Sales this year alone is projected to climb 7% to 1.3 million units.

Key components of HUD systems include mirrors, heat sinks, optical films (for light refraction), graphics processors, digital light-processing projectors, LCDs, LEDs, and OLEDs.

As the technology matures, HUDs are growing in size. The recent North American International Auto Show in Detroit displayed a system that projects an image 16 in. wide by 6 in. high, about twice the size of most current head-up images.

2.9.1.3 Requirements

Sensing Environment

Powertrain Control Module (PCM): Electronic signals from various sensors act like the engine's eyes and ears helping it make the most of its driving conditions.

Sensors are required for all the key functions necessary for

- Managingignition timing,
- Fuel delivery
- Emission control

- Transmission shifting
- Cruise control
- Engine torque reduction (if the vehicle has antilock brakes with traction control)
- Charging output of the alternator
- Controlling the throttle.

Reliable sensor inputs are an absolute must if the whole system is to operate smoothly.

The powertrain control module (PCM) is required to learn and make small adjustments to the fuel mixture and other functions over time as the vehicle accumulates miles.

In the case that the PCM controls the transmission, it may take a while to relearn the driver's habits, so the transmission may not shift in exactly the same manner as before until this occurs.

Remember, a PCM needs all its sensor inputs, proper battery voltage, a good ground, and the ability to send out control signals to function normally.

Autonomous Vehicle: "*Autonomous Vehicles*," that is, vehicles that still need a driver to take over the steering wheel or acceleration/braking pedal functions in case of unanticipated events.

Key Requirements

In addition to the requirements described for IoT Empower Vehicles, Autonomous vehicle will require safety-compliant MCU's and optical sensors.

Functional safety needs to be increasingly important for ASIL (Automotive Safety Integrity Level)-compliant MCUs.

- Processor units are predicted to reach almost half a billion dollars by 2020, from just $69 million last year.
- *Sensing Requirement*. Among the sensors used for autonomous driver assistance applications are optical sensors for navigation purposes.
- Driver assistance applications optical sensors are expected to grow sevenfold in the period spanning 2013–2020.

Requirement Considerations

Hybrid vehicles use discrete power and modules extensively in the engine management system, to control the motor/generator units either when running from the hybrid powertrain, or when energy is being stored under braking. Other considerations such as impact of next-generation wireless in the networking solution can be found in [23, 24].

Need for Insulated-Gate Bipolar Transistors (IGBTs)

IGBT is a high-voltage, high-current switch connected directly to the traction motor in a hybrid electric or electric vehicle. It takes direct current energy from the car's battery and, through the inverter, converts the alternating current control signals into the high-current, high-voltage energy needed to commutate or turn the motor. The IGBT is an ideal motor inverter switch for 20–120 KW EV motors due to its high efficiency and fast switching. The more efficient the IGBT, the less power is lost to wasted heat, resulting in better "mileage or miles per watt of energy."

Safety is a main concern for cars and their drivers; therefore, ISO 26262, an international functional safety standard for electrical and electronic systems in automobiles, was published in November 2011. Examples of automotive applications that must meet the standard include EV battery management, steering, braking, transmission, and powertrain. TI is a member of the ISO 26262 working groups and leads the semiconductor subgroup.

Need for a solution to include optimized power efficiency, ranging from high-power modules, microcontrollers, to sensors and discrete components. The following are the key areas:

- The first area is the main inverter, which controls the electric motor to determine driving behavior and captures kinetic energy released through regenerative breaking, feeding recovered energy back to the battery.
- The second is the DC/DC converter module, which supplies the 12-V power system from the high-voltage battery.
- The third area covers the auxiliary inverters/converters, which supply power on demand to systems such as air-conditioning, electronic power steering, oil pumps, and cooling pumps.
- The fourth area is the battery management system, which controls the battery state during charging and discharging to enable the longest possible battery life.
- The fifth area is the on-board charger unit, which allows the battery to be charged from a standard power outlet.

Other requirements include the following:

- Intelligent power switches for anti-lock brake systems
- MEMSs inertial sensors for automotive airbags
- Telematics microprocessor for General Motors' OnStar
- Increasing electronics integration for infotainment
- Near-field communications technology or NFC
- Longer driving range between charges
- Faster battery charging times

- Safety and security, car access
- In-vehicle networks
- Secure connected mobility
- Car-to-car communication
  - Car-to-infrastructure communication
  - Remote car management and broadcast reception.

Driverless Vehicle

A driverless vehicle, also defined as a robotic vehicle [25], is capable of sensing its environment and navigating without human input. Robotic cars exist mainly as prototypes and demonstration systems. Currently, the only self-driving vehicles that are commercially available are open-air shuttles for pedestrian zones that operate at 12.5 miles per hour (20.1 km/h). According to a recent study by IHS research, nearly 12 million self-driving cars are being sold annually and almost 54 million will be in use on global highways by 2035 [26].

Autonomous vehicles sense their surroundings with techniques such as radar, lidar, GPS, and computer vision. Advanced control systems interpret sensory information to identify appropriate navigation paths, as well as obstacles and relevant signage. Some autonomous vehicles update their maps based on sensory input, allowing the vehicles to keep track of their position even when conditions change or when they enter uncharted environments.

Unlike today's autonomous cars, Google's driverless car totally depends on its array of sensors, semiconductor ICs, and algorithms running inside several ECUs. ECUs are solely responsible for

- The safety of the passengers inside the driverless car
- The safety of pedestrians and other vehicles.

MCUs and other semiconductor ICs used in these ECUs need to be compliant with stringent safety certifications such as ISO 26262 or ASIL. ASIL-compliant chips cost more than the standard ICs.

Advanced electronics with higher computational capabilities and the absence of human intervention in driverless cars demands higher functionality – via algorithms – from an ECU. With this higher number of algorithms, the following trends will be evident:

- The number of cores and DMIPS in a processor chip will need to increase
- The use of Ethernet or FlexRay modules for higher bandwidth and secure communication
- An increase in the size of nonvolatile memory for storing huge amounts of data
- An increase in volatile memory to support image processing and to execute code.

Driverless, but not entirely autonomous, autonomous driver assistance systems have active-control mechanisms that take over the control of a car from the driver only to brake or steer in avoiding an accident when the driver does not respond to the warnings.

Adaptive cruise control (ACC) and Automatic emergency braking (AEB) are few examples. An autonomous car is a vehicle that not just spontaneously brakes or steers but drives a car automatically in different driving scenarios without human intervention. As shown in the picture below, Google's driverless car is still not "fully autonomous" because of limited operational conditions, especially with a speed limit of 25 miles per hour.

Requirement for Vehicle–Vehicle

Safety applications using V2V technology need to address a large majority of crashes involving two or more motor vehicles. With safety data such as speed and location flowing from nearby vehicles, vehicles can identify risks and provide drivers with warnings to avoid other vehicles in common crash types such as rear-end, lane change, and intersection crashes. These safety applications have been demonstrated with everyday drivers under both real-world and controlled test conditions.

The safety applications currently being developed provide warnings to drivers so that they can prevent imminent collisions, but do not automatically operate any vehicle systems, such as braking or steering. NHTSA is also considering future actions on active safety technologies that rely on on-board sensors. Those technologies are eventually expected to blend with the V2V technology. NHTSA issued an Interim Statement of Policy in 2013 explaining its approach to these various streams of innovation. In addition to enhancing safety, these future applications and technologies could help drivers to conserve fuel and save time.

2.9.2 Smart Roadside

2.9.2.1 Challenge

Increasing truck travel demand is resulting in too many legally loaded commercial motor vehicles queued up at inspection stations and, thus, unnecessary delays. Levels of enforcement are not keeping pace with this increase in trucks traveling – resources are being strained to deliver effective enforcement programs to ensure that all users of the highway are safe. The Smart Roadside will allow screening of trucks and drivers using wireless communication between the vehicle and the infrastructure while they travel at highway speeds.

2.9.2.2 Objectives

The objective of the smart roadside is the development and advancement of freight technology and operations by improving data sharing between industry and government. A second objective is overseeing enforcement of government size and weight limits by relying on law and enforcement agencies.

The smart roadside allows truck and driver to be screened with roadside sensors. Regulatory functions are employed while not interrupting the travel of compliant carriers. Sensors can provide shippers greater visibility of good movement.

2.9.2.3 Smart Roadside Requirements

- Standards and specifications are called for weigh-in-motion technology
- Development of virtual Weigh Station sites
- Development of Bridge Weigh-in-Motion Systems
- Test wireless roadside inspection
- E-Tolling
- Over-Height detector
- Weather monitoring station
- In-Vehicle monitoring
- Radiation detection systems
- Truck parking.

Smart trucks will be connected to smart roads. See Figure 2.23.

Technology options that can be used to identify electronically every commercial vehicle include

- Dedicated short-range communications (DSRCs)
- Other RFID such as windshield, license plate, or door-mounted placard

Figure 2.23 C-IoT and smart road for trucks

- Commercial Mobile Radio Service (CMRS)
- Optical readers.

2.9.2.4 Sensing Requirement

Smart systems/sensors are to be integrated with existing screening systems:

- Identification
- Dimension measurement (weight, height, width, and length)
- Smart infrared inspection system (tires and bearings)
- Radiation detectors
- Connectivity between trucks (V2V) and overall system.

Details can be found in dot.gov [27].

2.9.3 Drone

Drones Approved. FAA Gives OK to First Commercial Use Over Land (June 10, 2014)

- An AeroVironment Puma drone [28] undergoes preflight tests in Prudhoe Bay, Alaska, on Saturday, June 7.
- The drone will be used to survey roads, pipelines, and other equipment at the largest oil field in the United States.

2.9.3.1 Surveillance Drone

- Drones can carry various types of equipment including live-feed video cameras, infrared cameras, heat sensors, and radar.
- Mission planning software and tablet application streamline data transport and processing.
- They upload images and flight logs to the cloud server for fast data processing, analytics and access anywhere that can be viewed on mobile or desktop devices.

2.9.3.2 Privacy Concerns

- Drones carry Wi-Fi crackers and fake cell phone towers that can determine your location or intercept your texts and phone calls. Drone manufacturers even admit they are made to carry "less lethal" weapons such as rubber bullets.
- Concern of use among commercial establishments, hobbyists, and others.

Please see an example of a drone carrying a video surveillance camera in Figure 2.24.

Figure 2.24 C-IoT and drone

Considerations in designing Video Surveillance Networks.

- Processing and power management
- IPv6 for mobile devices
  - As more of the applications associated with video surveillance, such as sending feeds to mobile devices, operate over IPv6, there will be increasing momentum toward using IPv6 natively within video surveillance networks.
  - Governments are also beginning to mandate the use of IPv6 in public sector networking systems.

System Powering/Connectivity/Reliability/Security.

- Powering PoE + (Camera's power is delivered via data cable)
  - POE standard provides 15 W of PoE
  - Advanced camera's capabilities (zoom, pan, tilt) would require 30 W of power (PoE + standard)
- Use of the network's multicast signaling protocol to deliver video streams to multiple locations
- Redundancy
  - Double up on data paths
  - Ensure switch equipment support dual power supply units (PSUs)
  - Double up on power supply.

System Powering/Connectivity/Reliability/Security.

- Resilient backbone supporting multiple head-ends for data recording, network management, disaster recovery
- Bandwidth
  - Higher resolution images mean higher data rates. A 1 Gb uplink from a switch connecting 48 cameras might be enough today, but may well be inadequate within a few years.

  - Bandwidth should be provisioned to allow for up to a fivefold increase in bandwidth requirements within the installation's lifetime.
- Security
  - Configure high-security authentication on all camera-connected ports
  - Configure switches to send alarm messages if cameras are ever unplugged
  - Ensure that any switch ports to which cameras have not yet been attached are shut down
- Video Analytics
  - Create custom criteria-based alerts
  - Send alerts to you mobile device
  - Technology advancement to smart camera for facial recognition, license plate recognition, compare images.

2.9.4 Machine Vision

2.9.4.1 VTT – Embedded Industrial Solutions

- Machine vision systems for industry automation and microscope applications, to develop sensor prototypes with wireless data interfaces, DSP algorithms for signal extraction using multichannel spectral and statistical analysis, and hardware prototypes of embedded microprocessor systems
- Our expertise fields include developing and choosing suitable optics, illumination, cameras, frame grabbers, and software
- We are able to select, develop, and execute optimal image-processing algorithms
- We also have broad experience with various microcontrollers. We know how to build low-power embedded systems or cost-effective hardware-in-the-loop systems with fault injection capabilities.

Figure 2.25 shows an example of machine vision.

Figure 2.25 C-IoT and machine vision

2.9.4.2 Applications

- *Microscope Application* – Automatic quality control with machine vision for laser diodes
- Computer vision applications for recognition and tracking of owners unattended baggage in airports
- Data collection and analyzing system of driver's driving performance (where sensor seat foils are used to measure the driver's drowsiness)
- Sub-micron resolution in 3D Biomaterial Structuring
- Life science research in biophotonics.

2.9.5 *Smart City*

Focus on infrastructure, environment, and sustainability (air, water pollution monitoring) in bringing all the intelligence into a market.

According to Forrester Research, smart city is defined as a "city" that uses information and communications technologies to make the critical infrastructure components and services of a city – administration, education, healthcare, public safety, real estate, transportation, and utilities – more aware, interactive, and efficient. Information and Communications Technology (ICT) will play a key role in creating the foundation for smart cities – whether those cities are newer communities being built from scratch or centuries-old metropolises. Demand from local governments, along with similar conglomerations like universities and company towns, will drive incremental opportunities for ICT suppliers in the coming years [29].

2.9.5.1 C-IoT and Smart City

Of course, being able to control your home via a mobile device is the first step toward a truly "connected" home, and it is something already evident in the marketplace. But beyond mobile control both in-home and remotely, homes are being outfitted with smarter technology that can respond to outside stimulus and actively automate things like climate control and energy.

For example, the Advanced Homes technology allows the home to access meter energy data residing in the cloud and analyze it. The future of connected homes is really about a connected community; a connectivity that is beyond consumption of data by owners or controlling their lighting remotely.

As a starting point, municipalities have begun replacing the current light with LED lighting, which is more economical, reliable, and long lasting. This is due to the rapid advancement in LED technology, increase in production, and consequently decrease in LED pricing. Many estimates suggest that lighting as a whole accounts for about one-fifth of global electricity consumption, and LED-based lighting has had a major impact on that figure.

Figure 2.26 shows an example of what a smart city may consist of.

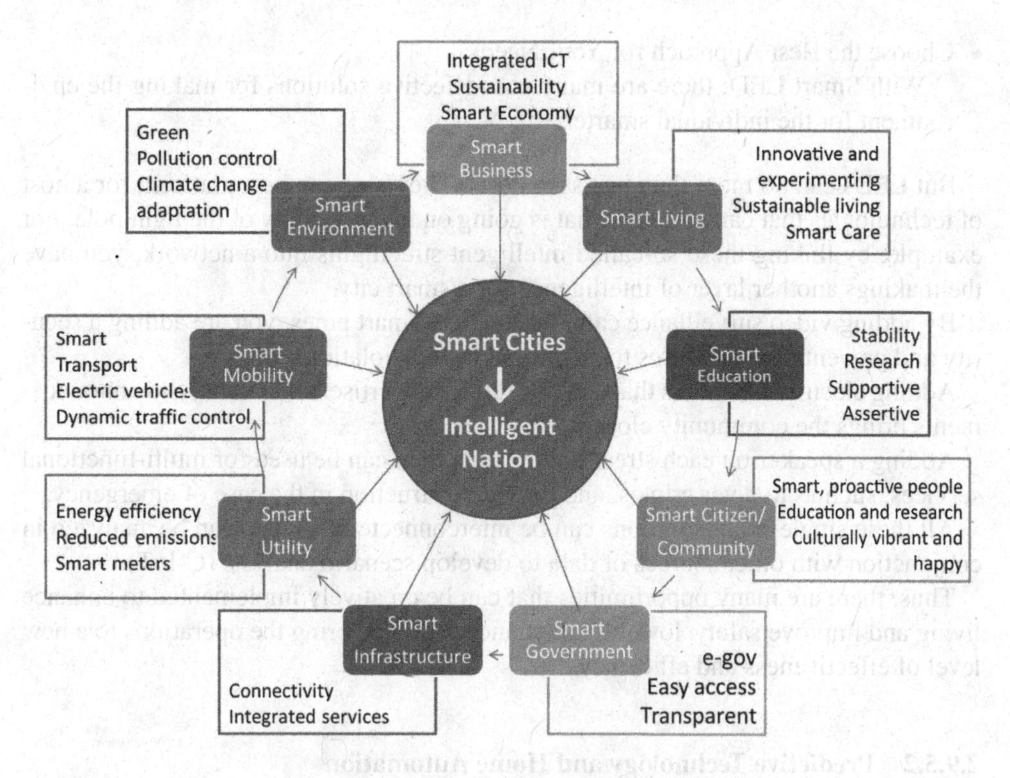

Figure 2.26 C-IoT and smart city

When Government/municipalities set direction for adopting LED technology, this will have a chain effect on the other domains, industries/utilities and individual/home.

Benefits

- Cities: Enhance public safety and accelerate environmental initiatives, while keeping costs low with increased operational efficiencies. Provide more valuable services to citizens and build regional economic advantage.
- Utilities:
  Grow value continuously from existing smart grid investments and streamline operations with integration across existing back-office systems. Eliminate the need for multiple networks and increase ROI.
- Simple Management:
  Make quick adjustments to street lighting based on changes in weather, traffic, public events, or accidents. Implement controlled run time and automated dimming to achieve up to 40% in operational savings.
- Seamlessly Add Smart City Services:
  Easily integrate other smart city devices and applications, such as intelligent traffic signal control, networked parking meters, and environmental sensors.

- Choose the Best Approach for Your Needs:

  With Smart LED; there are many cost-effective solutions for making the environment for the individual smarter.

But LEDs can do more than just save costs. They also can be a platform for a host of technologies that can monitor what is going on in the vicinity of the light pole. For example, by linking these so-called intelligent streetlights into a network, you have the makings another layer of intelligence for a smart city.

By adding video surveillance cameras to street smart poles, you are adding a security and preventative measures to crime and traffic violation.

Adding electronic banners that can be used for advertisement, alerts, and announcements brings the community closer.

Adding a speaker on each street pole, which then can be used for multi-functional services, such as to deter crimes, and provide instruction in the case of emergency.

All these single-point solutions can be interconnected and data can be analyzed in conjunction with other sources of data to develop scenario analysis (C-IoT).

Thus, there are many opportunities that can be creatively implemented to enhance living and improve safety, lower maintenance cost, and bring the operations to a new level of effectiveness and efficiency.

2.9.5.2 Predictive Technology and Home Automation

Imagine a TV that turns itself on right before your favorite show starts, or based on your past habits, predictively mutes itself when the commercials begin. Imagine a refrigerator that sends you a friendly reminder to pick up more orange juice when you are on your way home from work (which it knows you are right now, based on your GPS location data). The home of the future is one that knows your behavior, and responds to your routine accordingly.

Beyond just learning the homeowner's preferences and routine, smart homes will be able to self-manage things like energy consumption and climate control. Taking into account the homeowner's long-term energy goals as well as things like outside weather and city resources based on the smart grid, this future home will actively optimize its activity in real time.

According to Forrester, a smart environment uses information and communications technologies to make the critical infrastructure components and services of a city administration, education, healthcare, public safety, real estate, transportation and utilities more aware, interactive, and efficient.

In our definition, we make the definition more user-centric and do not restrict it to any standard communication protocol. This will allow long-lasting applications to be developed and deployed using the available state-of-the-art protocols at any given point in time. Our definition of IoT for smart environments is the Interconnection of sensing and actuating devices providing the ability to share information across platforms through a unified framework, developing a common operating picture for enabling innovative applications. This is achieved by seamless large-scale sensing, data analytics, and information representation using cutting-edge ubiquitous sensing and cloud computing. This requires vision and innovation that will empower design and development of smart systems [30].

Each day, the world becomes more and more connected through the "IoT"– a phrase that describes the expanding role the Internet plays in everything we do. In fact, according to some estimates, smart, Internet-enabled devices are expected to grow to 50 billion by 2020. That means that virtually smart devices and Internet connections affect every industry across the globe. Whether it is on a personal level – an oven, washer and dryer, watch, phone, car, baby monitor, home security – or whether it is at a larger level like a healthcare network, a school district, or a law firm, we are connected now, more than ever.

Market, Drivers, Positioning

The technology that allows us to be connected on so many levels is astonishing. It is fundamentally changing our world. So many things can now be done faster, more effectively, more efficiently, and more conveniently. They save time, energy, and money, while allowing us to focus our efforts on what truly matters most to us.

Big data too is playing an important role in the technology revolution. And because of big data in the cloud, companies of all sizes can access huge amounts of data to improve products and services for consumers the world over.

Unfortunately, all the good that is coming about with these tech advances can be drastically diminished with one fell swoop of a cyber attack, and next time, it could be your car being hacked instead of your router or cloud data. While the efforts companies are making to bring the best technology to individuals and businesses is laudable, the security measures still have a long way to go.

The list of large-scale data breaches is getting bigger and bigger. Recently, we witnessed the hacking of data of millions of consumers of major stores like Target and Wal-Mart. Microsoft recently revealed that there is a severe flaw in its Internet Explorer browser and millions of consumers were affected by the recent the Heart bleed bug. The list is growing and cyber attacks will continue.

The effects of a cyber attack on one device or company are scary, but when you include a network with thousands of connected devices or a family that is totally connected, the fear worsens dramatically.

Hackers could potentially control drug infusion pumps, which control things like pain medicine delivery, chemotherapy, and antibiotics. Many of the medical assets like in X rays, defibrillators, and other vital systems are found to be vulnerable. A cyber thief could also potentially blue-screen or completely shut down equipment. Information being sent across the network could be changed and manipulated whereby physicians could misdiagnose a patient, or prescribe an unneeded medication.

Two of many things that these findings show case are (i) any industry can be vulnerable to cyber attack and (ii) consumers need to demand that vendors provide the best in data security before they buy a product.

Some of the findings show that the main responsibility for the equipment flaws was with the vendors themselves. Too much equipment did not require any authentication and many of the hard coded passwords were too simple like "admin" or "1234." And while much responsibility does lie with the vendors, companies too need to be aware of the dangers. Many hospitals were totally unaware of the risk and problems.

Technology companies are aware of the security threats IoT imposes. In fact, Cisco has launched the "Internet of Things Grand Security Challenge," offering up to $300 000 worth of prizes for Internet security innovation in hopes of finding a solution. While such efforts should certainly be lauded, it begs the question how far behind innovation in security will fall behind innovation in the IoT.

What does this all mean? Most importantly it means that individuals and companies need to take every necessary step to ensure their cyber safety. That safety cannot be taken for granted anymore. Fortunately, by taking the often times simple and necessary steps all entities can protect themselves from cyber attacks. As important as it is to take action, companies and people need to stay up to date with security advances, install the latest software, update passwords, implement new technologies, And do everything possible to keep their data and network safe.

2.9.5.3 Requirements

1. *Individual*. Smart shopping Home – for example, identify food items to buy, identify stores that have sales or coupons; perhaps, order online and have a smart car to pick up the items on being notified when they are ready
2. *Industry*. Smart retails, shopping malls, parking, transactions (epos),
3. *Infrastructure*. Smart parking, smart street lighting, digital signage, public safety, and disaster management.

References

[1] Weightless http://www.weightless.org/ (accessed 18 November 2014).
[2] Wikipedia http://en.wikipedia.org/wiki/Body_area_network (accessed 18 November 2014).
[3] Behmann, F. (2005) Impact of Wireless (Wi-Fi, WiMAX) on 3G and next generation – an initial assessment. IEEE Xplore, Elecro Information Technology, 2005 IEEE International Conference, Lincoln, NE, May 22-25, 2005, Print ISBN: 0-7803-9232-9, http://ieeexplore.ieee.org /xpl/login.jsp?tp=&arnumber=1626995&url=http%3A%2F%2Fieeexplore.ieee.org%2Fiel5% 2F10832%2F34150%2F01626995 (accessed 18 November 2014).
[4] Behmann, F. (2005) Business view of the convergence/integration of Wi-Fi, WiMAX and 3G cellular. IEEE Communications and Signal Processor Societies Joint Chapter, Austin, TX, June 16, 2005, http://www.embeddedstar.com/weblog/2010/09/01/power-3gpp-lte/.
[5] Behmann, F. (2010) Power Architecture Enabled Differentiated Solution for LTE. Embedded Systems, Sep 1, 2010, http://www.embeddedstar.com/weblog/2010/09/01/power-3gpp-lte/ (accessed 18 November 2014).
[6] Behmann, F. (2012) SOC architecture evolution as a key driver of wireless network transformation. International Conference on Communications, ICC 2012, Ottawa, Canada, June 10–15, 2012, Session SA03, www.ieee-icc.org/2012 (accessed 18 November 2014).
[7] ITRS (2007) ITRS, 2007 Edition, Executive Summary, pp. 8–9, http://www.itrs.net/Links/ 2007ITRS/Execsum2007.pdf (accessed 18 November 2014).

[8] ITRS-System Drivers (2007) ITRS, 2007 Edition, System Drivers, pp. 3–7, http://www.itrs.net/links/2007_ITRS_CD/2007_Chapters/2007_SystemDrivers.pdf (accessed 18 November 2014).
[9] Behmann, F. (2009) The ITRS Process Roadmap and Nextgen Embedded Multicore SoC Design, March 22, 2009, pp. 1–4, http://www.embedded.com/design/mcus-processors-and-socs/4008253/The-ITRS-process-roadmap-and-nextgen-embedded-multicore-SoC-design (accessed 18 November 2014).
[10] Behmann, F. (1980) Virtualization for Embedded Power Architecture. Electronic Product Magazine, pp. 32–33, http://electronicproducts.com SEPTEMBER 2009.
[11] Open Networking https://www.opennetworking.org/images/stories/downloads/sdn-resources/solution-briefs/sb-sdn-nvf-solution.pdf (accessed 18 November 2014).
[12] Computer Hope (2007) Metcalfe's Law. http://www.computerhope.com/jargon/m/metcalfe.htm (accessed 18 November 2014).
[13] Wireless Sensor Networks http://www.sensor-networks.org/ (accessed 18 November 2014).
[14] IETF RFC 4919, http://tools.ietf.org/html/rfc4919 (accessed 18 November 2014).
[15] Research2guidance http://research2guidance.com/r2g/mHealth-App-Developer-Economics-2014.pdf (accessed 18 November 2014).
[16] Behmann, F. (2014) IP video surveillance – an irreversible trend for the future. IEEE Communications & Signal Processing Joint Chapter, Austin, July 17, 2014, http://sites.ieee.org/ct-comsp/ (accessed 18 November 2014).
[17] Behmann, F. (2008) Wireless Technology and Applications Trends, Freescale FTF, June 16, 2008, Session AN142.
[18] Behmann, F. (2012) Developers … how to meet advanced SoC architecture requirements? DesignWest, San Jose, CA, March 26–29, 2012.
[19] Behmann, F. (2012) Power Architecture Combines Rich Features for Embedded. RTC Magazine (May 5, 2012), http://www.rtcmagazine.com/articles/view/102566 (accessed 18 November 2014).
[20] Modern Material Handling (2006) RFID Market to Reach $9.2 Billion in 2014 and More Than Triple in Next Decade, http://www.mmh.com/article/rfid_market_to_reach_9.2_billion_in_2014_and_more_than_triple_in_next_decad (accessed 18 November 2014).
[21] IFR http://www.ifr.org/service-robots/ (accessed 11 December 2014).
[22] Mouser http://www.mouser.com/applications/automotive-power-semiconductors/ (accessed 18 November 2014).
[23] Behmann, F. (2006) Convergence challenges and solutions for next-generation wireless networking. Keynote presentation at IEEE- MTT Wireless, Jan 15–20, 2006.
[24] Behmann, F. (2007) Non-automotive technologies for the automotive engineer – networking solution offering. Freescale NCSG Conference, June 25, 2007, Session AA331.
[25] Casact.org (2014) The Future of Driving – Assistive Technology; Autonomous Vehicles, March 21, 2014, http://www.casact.org/community/affiliates/canw/0314/Covington_AutonomousVehicles.pdf (accessed 18 November 2014).
[26] IHS.org (2010) Self-Driving Cars Moving into the Industry's Driver's Seat, http://press.ihs.com/press-release/automotive/self-driving-cars-moving-industrys-drivers-seat (accessed 18 November 2014).
[27] DOT.Gov SmartRoad http://www.its.dot.gov/presentations/pdf/Integrated_Truck_Program.pdf (accessed 18 November 2014).
[28] Avinc (2012) AeroVironment Puma Drone, http://www.avinc.com/ (accessed 18 November 2014).
[29] Forrester Research (2014) Forrester Research on Smart Cities, http://www.urenio.org/2010/12/04/forrester-research-on-smart-cities/ (accessed 18 November 2014).
[30] Behmann, F. (2011) Power architecture 20 years of innovation empowering design and developer community. CSIA-ICCAD Annual Conference and China IC Design Industry, Beijing, china, https://www.power.org/events/2011-csia-iccad-annual-conference-china-ic-design-industry-10-year-achievements-exhibition/ (accessed 18 November 2014).

3

C-IoT Applications and Services

In this chapter, we describe some Collaborative Internet of Things C-IoT applications and services that span multiple domains, from personal consumer to home to industrial (business) and smart city (infrastructure and communities) levels to deliver sustainable smart living and smart environment that help optimize business process efficiency and improve quality of life, see Figure 3.1.

1. Smart living (Consumer)
   (a) Smart connected consumer – tracking and fitness/health monitoring
   (b) Smart connected home
   (c) Smart connected car, transport
2. Smart industry (Business)
   (a) Smart industrial (factories, buildings, smart grid/energy, retail, manufacturing)
   (b) Smart agriculture
3. Smart infrastructure (city and communities) and sustainable smart environments.

Here are some quotes regarding the impacts of the IoT (Internet of Things) markets and services.

- *Home and Building Automation.* Digital marketer Lauren Fisher points to the Nest Learning thermostat, which takes data about the home environment and owners' temperature preferences and programs itself to operate efficiently within the context of that information. This technical framework provides energy providers with the connectivity to better manage the energy grid.
- *Smart Car.* Mobile virtual network operator Alex Brisbourne describes how the automotive industry is increasingly designing automated applications into vehicles to provide maintenance monitoring, fuel and mileage management, driver security, and other capabilities that cost little to integrate but have significant earning potential. The addition of a cloud-based server to analyze the data and automatically

Collaborative Internet of Things (C-IoT): For Future Smart Connected Life and Business, First Edition.
Fawzi Behmann and Kwok Wu.

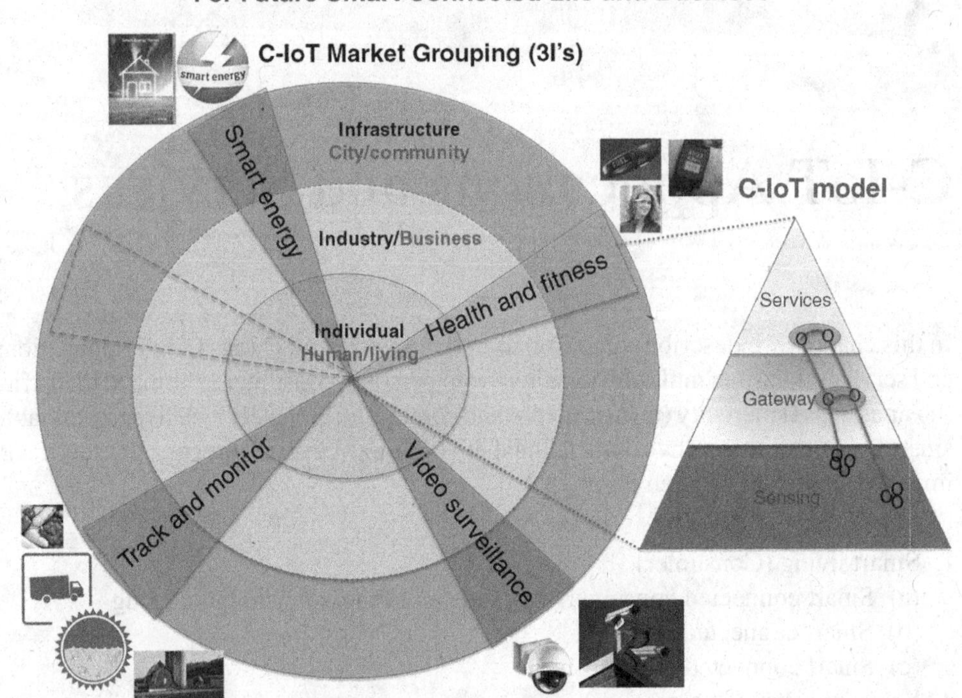

Figure 3.1 IoT markets

act on it – automatically scheduling a maintenance appointment at the appropriate time, for example – would move this further in the direction of the IoT.

- *Smart Transportation/Smart Cities.* Technology writer Martyn Casserly cites the London iBus system, which "… works with information from over 8000 buses that are fitted with global positioning system (GPS) capabilities alongside various other sensors which relay data about the vehicle's location and current progress," so bus stop signposts can display details of a bus's impending arrival.

IoT concepts have already been adopted in areas such as "energy (e.g., smart lighting and smart grid) and industrial automation … essentially whatever is connected to or crosses over the Internet." Cisco estimates the Internet of Everything (IoE) to be worth $14.4 trillion to the global economy by 2020.

3.1 Smart IoT Application Use Cases

This chapter explains several use case examples to demonstrate smart IoT systems with interoperability among smart devices to improve quality of life and business process efficiency [1, 2]. These IoT products for various IoT businesses can be deployed using a unified, secured, smart IoT software platform consisting of the IoT Gateway

Platform [3] and Sensor Fusion Platform [4] which will be described after the IoT use case examples [5].

There are three scalable domains (3I's) of influence with IoT products. Each IoT product for a certain business can span three domains.

The first domain, namely, Individual (Consumer) IoT level, Smart Connected Human living in a Smart connected Home will contribute toward improving quality of life in Smart Living. Applications from radio frequency identification (RFID) tracking of kids and pets to health fitness monitoring to smart thermostat in a smart home, all contribute to an improved lifestyle.

The second domain is Industrial IoT, from smart RFID in fleet tracking logistics to Smart Video security surveillance to active security systems with object recognition to smart energy meters to improve business process efficiency in Smart Business.

The third domain is the Infrastructure IoT for our city, community, and environment, to show how smart sensors operate together in our community infrastructure to improve our quality of life and sustainability of our smart city and smart world.

Many IoT business applications cross all three domains. Examples of these applications include video surveillance, smart energy, smart home/buildings, smart health, as well as tracking and monitoring of assets.

Motivation for C-IoT is the delivery of IoS (Internet of Service) (Figure 3.2). This chapter describes how embedded things can be connected to the cloud via the smart IoT gateway software platform for delivery of IoS business. Smart things (things with smart sensors and microcontrollers) are connected to the cloud (optionally via smart wireless gateways). IoS is an end-to-end system integration that delivers business services via the cloud, which utilizes a sensor-fusion software platform to perform data analytics and for decision making; automated actions and business services are thereby taken as part of streaming a business process. This referred to as B2C2T2B (Business to Consumer to Things to Business).

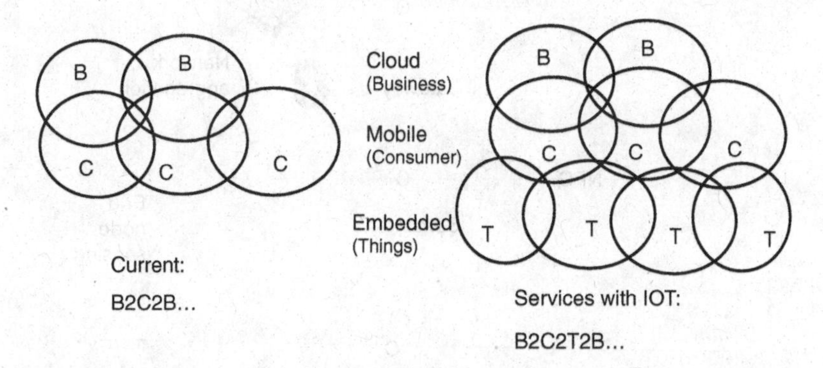

Figure 3.2 Internet of service (IOS) – integrate embedded things to cloud and mobile

3.1.1 Health Monitoring – Individual Level (Fitness/Health-Tracking Wearables)

These small consumer devices come in the form of smart bracelets, smart watches, smart eye-glasses, smart T-shirts, or smart shoes equipped with location sensors (RFID, near field communication (NFC), GPS) that track assets (kids, pets, elderly) as well as sensors for tracking health fitness biometrics (pulse, blood pressure, temperature, pedometer, accelerometer, etc.). The Go-Pro comes with a Wi-Fi camera mounted on a cap that capture actions using an MPU microprocessor unit to stream the video to smartphones/tablets which in turn can stream this to the cloud for sharing with friends. These devices are equipped with low-power microcontroller units (MCUs) which perform the data acquisition function and transmit the sensing data via Bluetooth® Low-Energy BLE (BT4.0) wireless connectivity to a smartphone or PC which ultimately transmits the sensing data onto the cloud for Big Data analytics and storage, thus allowing remote monitoring and tracking control.

The smart wearable fitness in Figure 3.3 shows an activity monitoring bracelet that contains an accelerometer and transmits its data to an intermediate gateway such as a smartphone or PC via Bluetooth BLE (BT4.0). The data are then transmitted to the Internet server for Big Data analytics. The summary trend analysis report is then accessible via the individual's smartphone. Table 3.1 shows software stacks for smart wearable health fitness/monitoring.

The wearable device usually pairs with the smartphone/smart tablet to act as an IoT gateway that aggregates the sensing data and transmits them to the cloud.

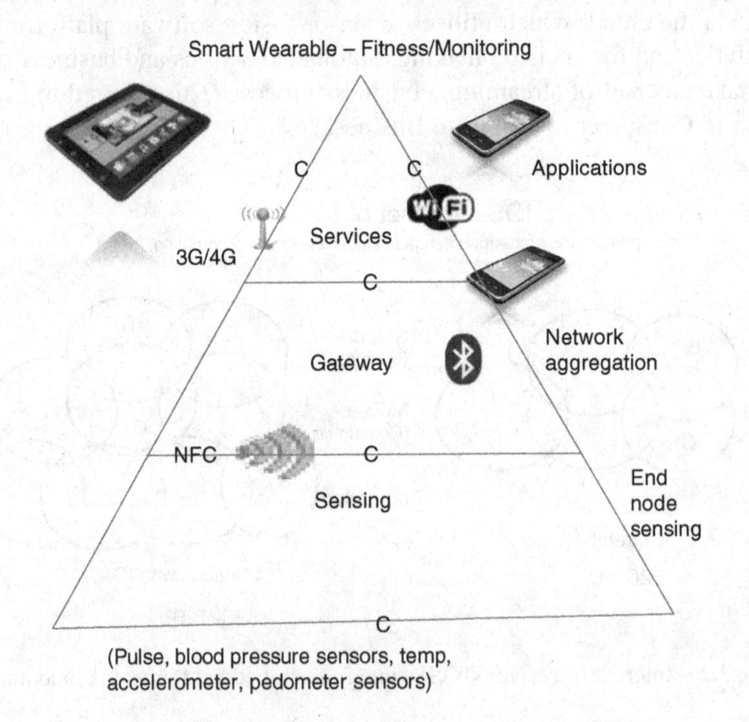

Figure 3.3 Heath fitness monitoring

Table 3.1 Software stacks for heath fitness monitoring

Service (cloud)	Networking ipv6, VLAN, DNS	DB – SQL, unstructured DB Cloud apps – billing, and so on				Ethernet driver		Web-server Storage
IoT gateway (wireless router)	Networking ipv6, VLAN, DNS	DB – SQL, unstructured DB	Wireless security, WPA2, WEP	BT driver	WiFi driver 3G/4G driver NFC driver	Ethernet driver	Other drivers, USB, SDIO, and so on	Storage
Sensing node			Wireless security, WPA2, WEP	BT driver	WiFi driver 3G/4G driver NFC driver	Ethernet driver	Other drivers, USB, SDIO, and so on	

A Machine-to-Machine (M2M) Gateway (sometimes referred to as *data aggregator* or *concentrator*) is used to provide connectivity among multiples sensor end nodes and multiple users. An IoT Gateway is essentially an M2M Gateway with added cloud (Internet network) connection that facilitates Big Data analytics and remote monitoring/control.

A fitness activity monitoring bracelet can also connect to other fitness devices such as a smart weight scale that monitors body weight, body fat, so the consumer's weight and fat are connected to the fitness monitoring database for more accurate computing of factors such as calorie consumption.

Another form of wearable is the smart watch, which can include camera, accelerometer, thermometer, altimeter, barometer, compass, chronograph, calculator, cell phone, touch screen, GPS navigation, Map display, graphical display, speaker, scheduler, and watch functions. A smart watch usually has NFC functionality that can pair with a smartphone for data exchange synchronization with the smartphone.

When these portable devices are equipped with RFID, GPS, and so on, they become useful for tracking and monitoring of assets in addition to fitness monitoring. Location tracking can also be implemented using BLE and/or low-power Wi-Fi for lower power, long battery life operation. Tracking of assets can be done at consumer level (kids, pets, elderly) at business and industry level (goods and cargo), and also in transportation logistics (taxis, trucks) at industrial and city infrastructure level. Asset monitoring and tracking can also be applied to agriculture such as to monitor the health of livestock and vegetation which includes automation of irrigation, feeding, and fertilization.

A smart wearable device can be evaluated by the following 8A's:

> 8A's: Automated Remote Provisioning and Management, Augmented Reality Human–Machine Interface HMI, Awareness of Context and Location, Analyze and Take Action, Automate, Anticipate, and Predict, Autonomous, Attractive

- *Automated Install.* This is usually a USB-based auto-install using a PC.
- *Augmented Reality and HMI (human–machine interface).* This usually has limited user-interface and display.
- *Awareness of Context and Location.* They usually lack awareness of context and location.
- *Analyze and Take Action.* Smart wearables usually have limited analysis capability and usually transmit sensing data to the Internet (cloud) for analysis.
- *Automate.* They have limited automation capability.
- *Anticipate and Predict.* Again, this is usually based on cloud-based analysis of data trend with limited predictive analysis capability.
- *Autonomous.* They involve limited rule-based intelligence and do not collaborate with other smart devices in the environment.
- *Attractive and Esthetic.* Smart wearables usually have good aesthetics, look, and feel.

3.1.1.1 Next Generation Smart Wearable and Automation Devices

Future Smart IoT systems will be more interoperable to create more conscious and thoughtful home operation and take connected intelligence to the next level of collective intelligence.

The smart wearable will interoperate with the smart thermostat to offer further context aware operation and take connected intelligence to the next level of collective intelligence.

For example, a fitness wristband's motion sensor could be used to detect that the owner is awake and trigger the smart thermostat to turn on the heating system.

The smart video camera could provide the smart thermostat a customized context control regarding who is at home.

3.1.2 Health Monitoring at Business Level (e.g., Clinic and Homes for the Elderly)

3.1.2.1 M2M in Healthcare

Within an aging population, an increase in monitored illnesses such as diabetes and heart disease and insurance mandates around hospital stays and visits has led to an increase in home-based health monitoring [6]. This is now being matched by the onset of portable devices, which monitor patients away from a hospital or physician's office. Cost savings match the added comfort of the patient for healthcare providers and insurers. Devices that monitor a patient's vital signs at home can operate as a direct M2M device via a gateway of the type mentioned in the previous M2M at Home section, or a dedicated telehealth hub. In either case, measured data such as blood pressure, heart rate, body temperature, respiratory rate, blood glucose, and cholesterol can be accumulated, processed and, if desired, sent periodically to the healthcare provider. Numerous MPU microprocessors are enabled with trust architecture and encryption acceleration hardware which help provide a secure encrypted communication link between patient and physician ensuring that private information does not get stolen.

In the IoT, devices gather and share information directly with each other and connect to the cloud, making it possible to collect, record, and analyze new data streams faster and more accurately. That suggests all sorts of interesting possibilities across a range of industries: cars that sense wear and tear and self-schedule maintenance or trains that dynamically calculate and report projected arrival times to waiting passengers.

But nowhere does the IoT offer greater promise than in the field of healthcare, where its principles are already being applied to improve access to care, increase the quality of care and most importantly reduce the cost of care. A telehealth product delivers care to people in remote locations and monitoring systems that provide a continuous stream of accurate data for better care decisions.

As the technology for collecting, analyzing, and transmitting data in the IoT continues to mature, we will see more and more exciting new C-IoT-driven healthcare applications and systems emerge. Read on to learn what is happening now – and what is on the horizon – for healthcare in the age of the IoT.

There is no shortage of predictions about how C-IoT is going to revolutionize healthcare by dramatically lowering costs and improving quality. Wireless sensor-based systems are at work today, gathering patient medical data that was never before available for analysis and delivering care to people for whom care was not previously accessible. In these ways, C-IoT-driven systems are making it possible to radically reduce costs and improve health by increasing the availability and quality of care.

An IoT-driven healthcare monitoring system includes

- Sensors that collect patient data
- Microcontrollers that analyze and wirelessly communicate the data
- Microprocessors that enable rich graphical user interfaces (GUIs)
- Healthcare gateways-analyzed sensor data that are sent to the cloud.

3.1.2.2 Understanding C-IoT

C-IoT-related healthcare systems today are based on the essential definition of the IoT as a network of devices that connect directly with each other to capture and share vital data through a secure socket layer (SSL) that connects to a central command and control server in the cloud. Let us begin with a closer look at what that entails and what it suggests for the way people collect, record, and analyze data – not just in healthcare, but in virtually every industry today.

The idea of devices connecting directly with each other is, as the man who coined the term Internet of Things puts it, "a big deal." As Kevin Ashton explained a decade after first using the phrase at a business presentation in 1999, "Today computers – and therefore, the Internet – are almost wholly dependent on human beings for information. The problem is, people have limited, time, attention, and accuracy – all of which means they are not very good at capturing data about things in the real world." The solution, he has always believed, is empowering devices to gather information on their own, without human intervention.

The following are two important reasons for devices to connect directly to data and to each other:

1. Advances in sensor and connectivity technology are allowing devices to collect, record, and analyze data that was not accessible before. In healthcare, this means being able to collect patient data over time that can be used to help enable preventive care, allow prompt diagnosis of acute complications, and promote understanding of how a therapy (usually pharmacological) is helping improve a patient's parameters.
2. The ability of devices to gather data on their own removes the limitations of human-entered data – automatically obtaining the data physicians need, at the

time and in the way they need it. The automation reduces the risk of error. Fewer errors can mean increased efficiency, lower costs, and improvements in quality in just about any industry.

3.1.2.3 C-IoT in Action in Healthcare

C-IoT plays a significant role in a broad range of healthcare applications, from managing chronic diseases at one end of the spectrum to preventing disease at the other. Here are some examples of how its potential is already playing out:

Clinical Care. Hospitalized patients whose physiological status requires close attention can be constantly monitored using C-IoT-driven, noninvasive monitoring. This type of solution employs sensors to collect comprehensive physiological information and uses gateways and the cloud to analyze and store the information and then send the analyzed data wirelessly to caregivers for further analysis and review. It replaces the process of having a health professional come by at regular intervals to check the patient's vital signs, instead providing a continuous automated flow of information. In this way, it simultaneously improves the quality of care through constant attention and lowers the cost of care by eliminating the need for a caregiver to actively engage in data collection and analysis.

An example of this type of system is the Massimo Radical-7®, a health monitor for clinical environments that collects patient data and wirelessly transmits for ongoing display or for notification purposes. The results provide a complete, detailed picture of patient status for clinicians to review wherever they may be. The monitor incorporates an embedded processor with enhanced graphics capabilities that enables extremely high-resolution display of information, as well as a touch-based user interface (UI) that makes the technology easy to use.

3.1.2.4 Remote Patient Monitoring

There are people all over the world whose health may suffer because they do not have ready access to effective health monitoring, see Figure 3.4. But small, powerful wireless solutions connected through C-IoT are now making it possible for monitoring to come to these patients instead of vice versa. These solutions can be used to securely capture patient health data from a variety of sensors, apply complex algorithms to analyze the data and then share these through wireless connectivity with medical professionals who can make proper health recommendations.

As a result, patients with chronic diseases may be less likely to develop complications, and acute complications may be diagnosed earlier than they would be otherwise. For example, patients suffering from cardiovascular diseases who are being treated with digitalis could be monitored around the clock to prevent drug intoxication.

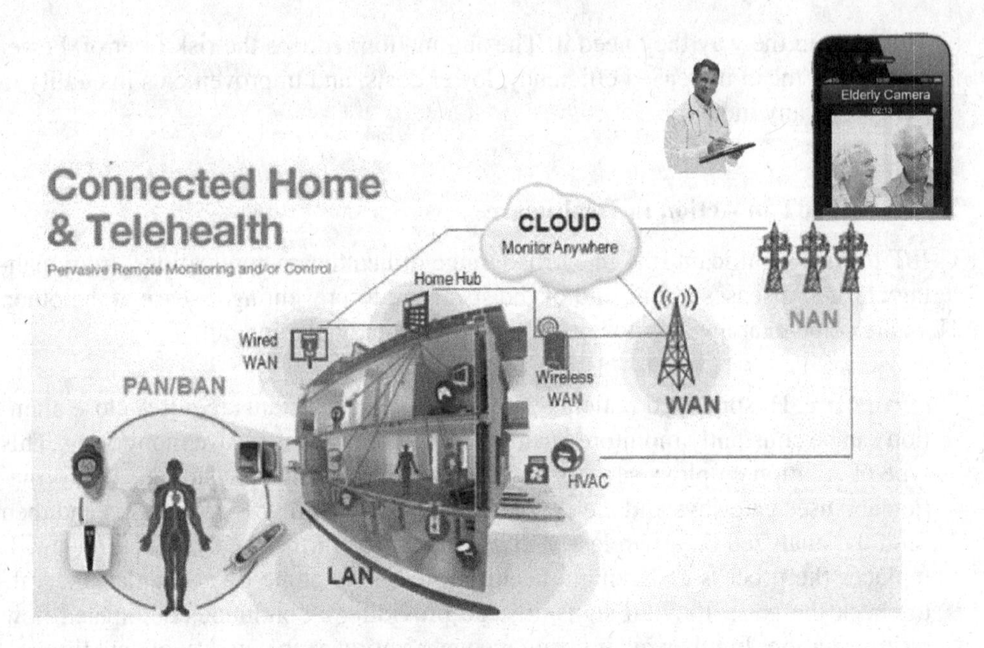

Figure 3.4 Remote patient monitoring

Arrhythmias that are randomly seen on an EKG could be easily detected, and EKG data indicating heart hypoxemia could lead to faster detection of cardiac issues. The data collected may also enable a more preventive approach to healthcare by providing information for people to make healthier choices.

An example of an enabling technology for remote monitoring is a Home Health Hub (HHH) Gateway reference platform, (Figure 3.5) built on an embedded processor, integrating with wireless connectivity and power management – in the telehealth gateway that enables collection and sharing of physiological information. The hub captures patient data from a variety of sensors and securely stores it in the cloud, where it can be accessed by those engaged in the patient's care. Data aggregation devices like this will soon become commonplace and will not only collect healthcare data but also manage other sensor networks within the home. In addition to healthcare data, this gateway manages data from smart energy, consumer electronics, and home automation and security systems.

Early intervention/prevention: Healthy, active people can also benefit from C-IoT-driven monitoring of their daily activities and well-being. A senior living alone, for example, may want to have a monitoring device that can detect a fall or other interruption in everyday activity and report it to emergency responders or family members. For that matter, an active athlete such as a hiker or biker could benefit from such a solution at any age, particularly if it is available as a piece of wearable technology.

These are just a few examples of C-IoT-based healthcare solutions, and many more are emerging. But as one reporter has noted, "The real vision for the future is that

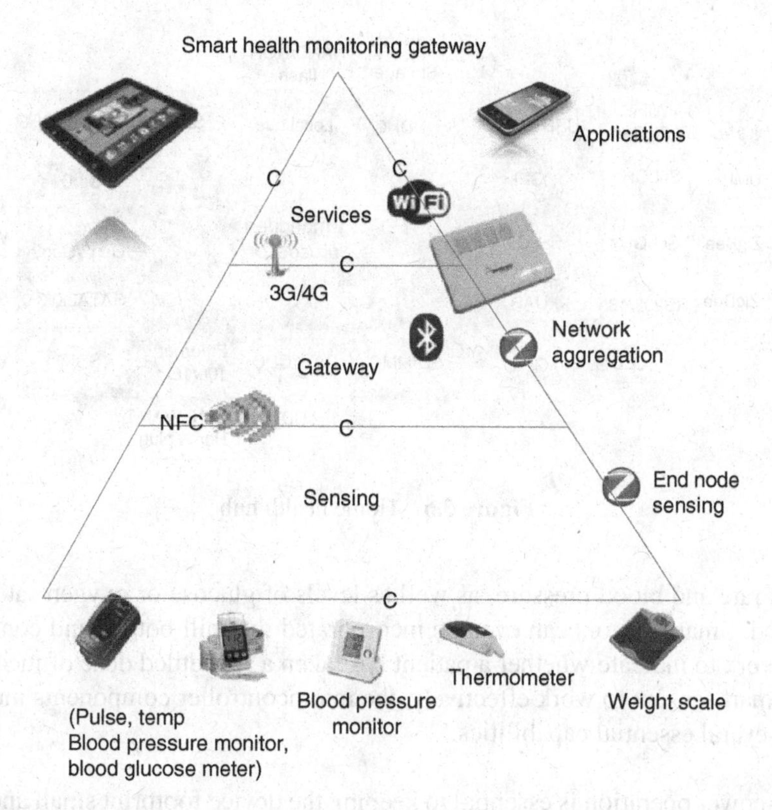

Figure 3.5 Smart health monitoring platform (e.g., clinic and elderly homes)

these various smaller applications will converge to form a whole.... Imagine if you are a relative of [a] patient who forgot their medicine. You receive the alert, are able to know their location, check their vital signs remotely to see if they are falling ill, then be informed by your car's navigation system which hospital has the most free beds, the clearest traffic route to get there and even where you can park."

3.1.2.5 Enabling Technologies: Making C-IoT in Healthcare Possible

The successful use of C-IoT in the preceding healthcare examples relies on several enabling technologies. Without these, it would be impossible to achieve the usability, connectivity, and capabilities required for applications in areas such as health monitoring.

Figure 3.6 shows a Smart Health Monitoring Platform used for clinics or in homes for the elderly.

Smart sensors, which combine a sensor and a microcontroller, make it possible to harness the power of C-IoT for healthcare by accurately measuring, monitoring, and analyzing a variety of health status indicators. These can include basic vital signs such

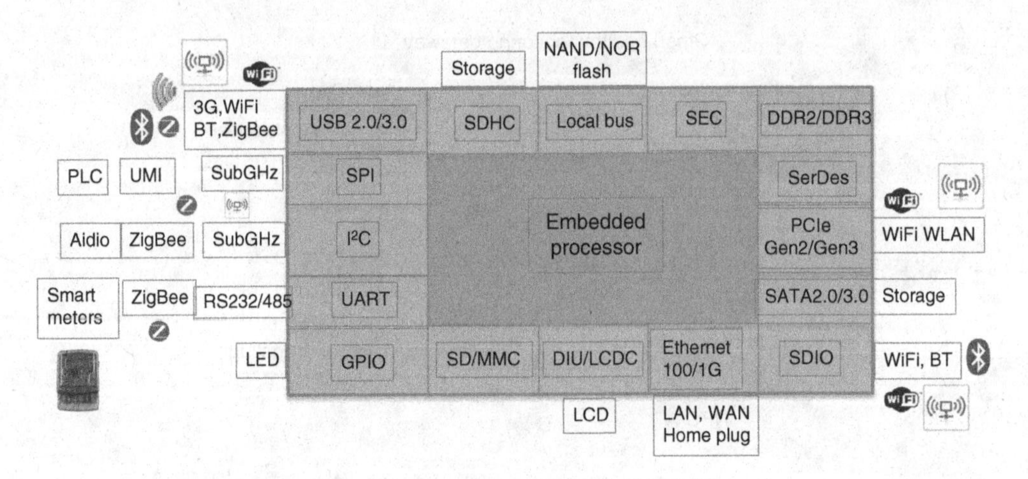

Figure 3.6 Home health hub

as heart rate and blood pressure, as well as levels of glucose or oxygen saturation in the blood. Smart sensors can even be incorporated into pill bottles and connected to the network to indicate whether a patient has taken a scheduled dose of medication.

For smart sensors to work effectively, the microcontroller components must incorporate several essential capabilities:

- Low-power operation is essential to keeping the device footprint small and extending battery life, characteristics that help make IoT devices as usable as possible. In the future, there will be more low-power processors that will be battery-free devices that utilize energy harvesting techniques through the use of ultra-low-power DC–DC (direct current) converters.
- Integrated precision-analog capabilities make it possible for sensors to achieve high accuracy at a low cost by selecting microcontrollers with high-resolution analog-to-digital converters (ADCs) and low-power op-amps.
- GUIs improve usability by enabling display devices to deliver a great deal of information in vivid detail and by making it easy to access that information.

Gateways are the information hubs that collect sensor data, analyze and then communicate the data to the cloud via wide area network (WAN) technologies. Gateways can be designed for clinical or home settings; in the latter, they may be part of larger connectivity resource that also manages energy, entertainment, and other systems in the home. Medical device designers can also use the platform to create remote-access devices for remote monitoring.

Wireless networking removes the physical limitations on networking imposed by traditional wired solutions such as the Ethernet and USB. MCUs and MPUs that support wireless connectivity for devices based on popular wireless standards such as Bluetooth® and BLE for personal area networks (PANs) are used with personal devices and Wi-Fi® and Bluetooth for local area networks (LANs) in clinics or hospitals. That leads us to a key challenge for the IoT in healthcare: standards.

3.1.2.6 Connectivity Standards: Enabling C-IoT Devices to Work Together

Standards represent an inherent challenge for any environment in which a large number of complex devices need to communicate with each other – which is exactly the case for C-IoT in healthcare. One analyst has described the "... greater standardization of communications protocols ... " as critical to advancing the adoption of C-IoT.

Fortunately, standards organizations are working now to create guidelines for wireless communications between monitoring devices and with care providers. The Continua Health Alliance is a coalition of healthcare and technology companies that was founded in 2006 to establish guidelines for interoperable personal health solutions. The organization has already published a set of specifications to help ensure interoperability. In the future, organizations that buy a Continua Certified® device will have the assurance that it will connect with other certified devices in IoT-driven applications. Continua's device standards are part of a larger standards environment that includes information technology standards established by the International Organization for Standardization (ISO) and engineering standards set by the Institute of Electrical and Electronics Engineers (IEEE®).

In wireless technology, IEEE standards for LANs define Wi-Fi (IEEE 802.11) and ZigBee® (IEEE 802.15.4) networks. Standards for PANs include Bluetooth and BLE, as well as IEEE 802.15.4j and IEEE 802.15.6, which are the IEEE standards associated with the body area network (BAN). Standards for cellular networks include GSM/UMTS and CDMA. Proprietary wireless networks still play something of a role in healthcare environments in general and IoT applications in particular, but that role seems to be shrinking as the industry continues to move toward standards-based architectures.

3.1.2.7 C-IoT in Healthcare

The long-predicted C-IoT revolution in healthcare is already underway, as the examples in this chapter make clear. And, those are just the tips of the proverbial iceberg, as new use cases continue to emerge to address the urgent need for affordable, accessible care. Meanwhile, we are seeing the C-IoT building blocks of automation and M2M communication continue to be established, with the addition of the service layer completing the infrastructure.

Table 3.2 shows the software stacks for a Health Monitoring Platform used for clinics or elderly homes.

Table 3.2 Smart health monitoring platform

		DB – SQL, unstructured DB						Web-server
Service (cloud)	Networking ipv6, VLAN, DNS	Cloud apps – billing, and so on				Ethernet driver		Storage
		DB – SQL, unstructured DB	Security, IPSEC, SSL, DTLS, encryption secured boot	NAT, ACL, fire-wall	HTTPS	OpenVPN, OpenSSL		Web-server
IoT gateway (wireless router)	Networking, ipv6, VLAN, DNS	Provisioning and Mgt TR69, QOS	Wireless security, WPA2, WEP					
				BT driver, ZigBee driver	WiFi driver, 3G/4G driver, NFC driver	Ethernet driver	Other drivers, USB, SDIO, and so on	Storage
							Other drivers, USB, SDIO, and so on	
	ZigBee driver		Security, IPSEC, SSL, DTLS, encryption, secured boot	BT driver	WiFi driver, 3G/4G driver, NFC driver	Ethernet, driver		
Sensing node			Wireless security, WPA2, WEP					

3.1.2.8 Home Health Hub (HHH)

The HHH (Figure 3.6) reference platform provides seamless connectivity with commercially available wired and wireless healthcare devices, such as blood pressure monitors, pulse oximeters, weight scales, and blood glucose monitors [7]. The data obtained from these devices is then relayed via Wi-Fi and 3G Broadband to a remote device, such as a smartphone, tablet, or PC, in order to track and monitor a patient's health status as well as provide alerts and medication reminders.

The display interface also provides a real-time connection to caregivers, including family, friends, and physicians, to bring peace of mind and safety to the person being monitored.

The HHH reference platform features a low-power embedded processor, a ZigBee® transceiver, and a sub-gigahertz radio transceiver used for a panic alarm sensor, providing personal emergency response system (PERS) functionality and enabling remote healthcare device monitoring.

3.1.2.9 Features Benefits

- The development and demonstration platform includes a gateway and a panic alarm sensor.
- The platform supports rapid prototyping, reducing time to market, and focusing resources on differentiation.
- Prevalidated USB, BT, BLE, Wi-Fi, ZigBee connectivity including medical class grades are available.
- The platform allows for connectivity to medical devices and sensors for automatic reporting and monitoring of vital sign measurements and implementation of daily activity alarms, and alarms for early detection of injury or security risks.
- Connectivity is available via Wi-Fi and Ethernet to external smart devices (tablet, smartphone, PC) along with a compelling UI for remote display.
- The platform offers anytime access and consultation to trusted health resources, medical staff, and family and friends through an intuitive and simple to use interface.

3.1.2.10 HHH Reference Platform Kit Contents

- HHH gateway printed circuit board (PCB)
- BT/Wi-Fi module (connects to the HHH Gateway PCB)
- Panic alarm sensor
- Quick start guide
- Windows Embedded Compact 7, Linux, Java frameworks with example code
- ZigBee Healthcare and Home Automation stacks
- Bluetooth HDP and Low-Energy stacks (subject to license from Stonestreet One)
- USB PHDC stack

- Wi-Fi stack
- Design files
- Cables.

3.1.3 Home and Building Automation – Individual Level (Smart Home)

3.1.3.1 Smart Thermostat (Smart Energy Management)

A smart thermostat replaces the traditional digital thermostat that has a fixed program by having added cloud connectivity for remote provisioning and updates, and by supporting remote monitoring and control via smartphones/tablets. A smart thermostat in the future could also act as a wireless gateway that interconnects other personal and home automation devices through a ZigBee-based wireless sensor network (WSN).

Here is an evaluation of Nest's Smart Thermostat with respect to the 8A's:

> 8A's: Automated Remote Provisioning and Management, Augmented Reality Human–Machine Interface HMI, Awareness of Context and Location, Analyze and Take Action, Automate, Anticipate, and Predict, Autonomous, Attractive

Automated Provisioning. The Nest Smart Thermostat is simple, intuitive, and easy to use, as it has a self-learning install and adaptive setup mode; it learns your preferred temperature settings during weekdays and weekends with its auto-schedule mode. One aspect of the self-learning is keeping track of Time-to-Temp, whereby it learns how long it takes your home to heat up and cool down, so it will get ready ahead of time before the present time of desired temperature. It will turn off the furnace but leave the fan on long enough to maximize heat distribution without wasting energy. This smart device is Internet connected.

- *Automated Updates*. It receives automated software updates as it becomes more intelligent. The cloud connection provides remote monitoring and controlability through smartphones/tablets.
- *Automated*. It is automated with multiple sensors such as temperature, humidity, ambient light, infrared motion, proximity short range, and long-range activity sensors.
- *Analyze and Take Action.* The humidity sensor can trigger humidifier to turn on, as the air starts getting dry.
- *Aware of Context and Location.* The auto-away mode is context aware with a motion sensor that can detect if people are around, and avoid wasting energy heating or cooling an empty house. It also leverages the Internet location-aware weather condition outside the home and customizes the heating and cooling accordingly.
- *Anticipate and Predict: Autonomous Action.* It is interoperable with smartphones/tablets and other devices such as the Nest Protect smoke alarm. It has an

auto-tune mode that automatically makes adjustment to lower energy consumption while keeping you comfortable. For example, its airwave mode automatically runs the alternating current (AC) less when humidity is not too high and ensures that you stay cool. It is provided with a "filter reminder" to remind you of the time for preventive care to replace the air filter. It could also remind you when to perform an AC tune-up.

- *Attractive.* It is easy on the eyes, with a stainless steel benzyl that reflects the surrounding wall color. A glass LCD display shows feedback regarding operating modes (red for heating) and efficiency (green leaf when it is saving energy).

  The round benzyl is a scroll wheel that can be turned like a trackball and the magneto sensor provides accurate menu location selection.

  This smart device is secured with WEP, WPA2, HTTPS, SSL, and 128-bit encryption.

Next-Generation Smart Thermostat and Home Automation Devices

Future smart C-IoT systems will be more interoperable to create more conscious and thoughtful home operation and take connected intelligence to the next level of collective intelligence.

For example, a Fitbit or Jawbone fitness wristband's motion sensor could be used to detect that the owner is awake and trigger the smart thermostat to turn on the heating system. When you leave home, your smart garage door openers could trigger your smart thermostat to a lower setting once you have left. As your car approaches home, your car can trigger the smart thermostat to turn on the heating, ventilation, and air-conditioning (HVAC). Your smart smoke alarm could trigger your LED lighting to flicker on lighting in addition to just the alarm. Your smart thermostat, while in the auto-away mode, can randomly turn on/off lighting while you are away from home.

The smart thermostat will interoperate with the smart IP camera to offer further context-aware operation and take connected intelligence to the next level of collective intelligence.

The smart video camera such as the "Dropcam" for surveillance could provide the smart thermostat a customized context control regarding who is at home.

3.1.3.2 Smart Smoke Alarm (Safety)

The Nest Protect smoke alarm is another smart home device from Nest Lab. It is equipped with a photoelectric smoke sensor, a carbon monoxide sensor, a heat sensor, activity sensors, a humidity sensor, and an ambient light sensor. The Nest Protect smoke alarm can interoperate with a Nest thermostat. Nest Protect uses a lower-power 100 MHz Cortex M3. A şmart smoke alarm also has multiple LED light display feedback, such as white light to indicate automatic night light, green to indicate all is clear, yellow to indicate early warning, and red for emergency and evacuation. The motion detector is used to detect directed arm-waving to silence the alarm.

3.1.3.3 Smart IP Camera for Video Surveillance (Security)

The "Dropcam" wireless IP camera, an example for video monitoring at homes and small businesses, is a dual-band (2.4 and 5 GHz) wireless IP camera with cloud-based SSL-encrypted video recording service using AWS (Amazon Web Service), which now records more videos than YouTube. This is an HD720p camera with night vision, 8× zoom, 130° viewing, two-way talk (mic and speaker), smart alerts (activity recognition based on motion and audio), and location aware.

Here is an evaluation of Dropcam with respect to the 8A's:

> 8A's: Automated Remote Provisioning and Management, Augmented Reality Human-Machine Interface HMI, Awareness of Context and Location, Analyze and Take Action, Automate, Anticipate, and Predict, Autonomous, Attractive

- *Automated Provisioning.* The Wi-Fi IP-camera has an easy setup wizard, Bluetooth (BT) pairing connectivity is provided, so one can even install the IP-camera via a smartphone.
- *Automated.* Smart alerts – activity recognition based on motion and audio.
- *Analyze and Take Action.* The Dropcam Pro has a pattern recognition video analytic feature to track custom rules. For example, you might set the tab to send an alert to your mobile phone when the front door is opened or when the TV or the desktop PC is moved.
- *Aware of Context and Location.* Turns camera on and off depending on where the owner is located and has the optional motion detect sensor "Dropcam Tab" which can be placed at a window or door within 100′ from the Dropcam camera.
- *Secure.* The video is SSL (Secured socket layer)-encrypted, and one can sign up for cloud-based video recording service.
- *Attractive.* It is easy on the eyes.

Next-Generation Smart IP Surveillance Camera and Home/Building Automation Devices

The Smart IP camera will interoperate with the Smart Thermostat to offer further context-aware operation and take connected intelligence to the next level of collective intelligence.

Your smart video camera could provide the Smart Thermostat a customized context control regarding who is at home.

Video monitoring is used in home and public security in retail, banks, ATMs, school, traffic monitoring, transport safety, as well as in factory and manufacturing automation such as machine vision in robotics for automated assembly and automated assembly inspection. Smart video analytic software is added for smart video

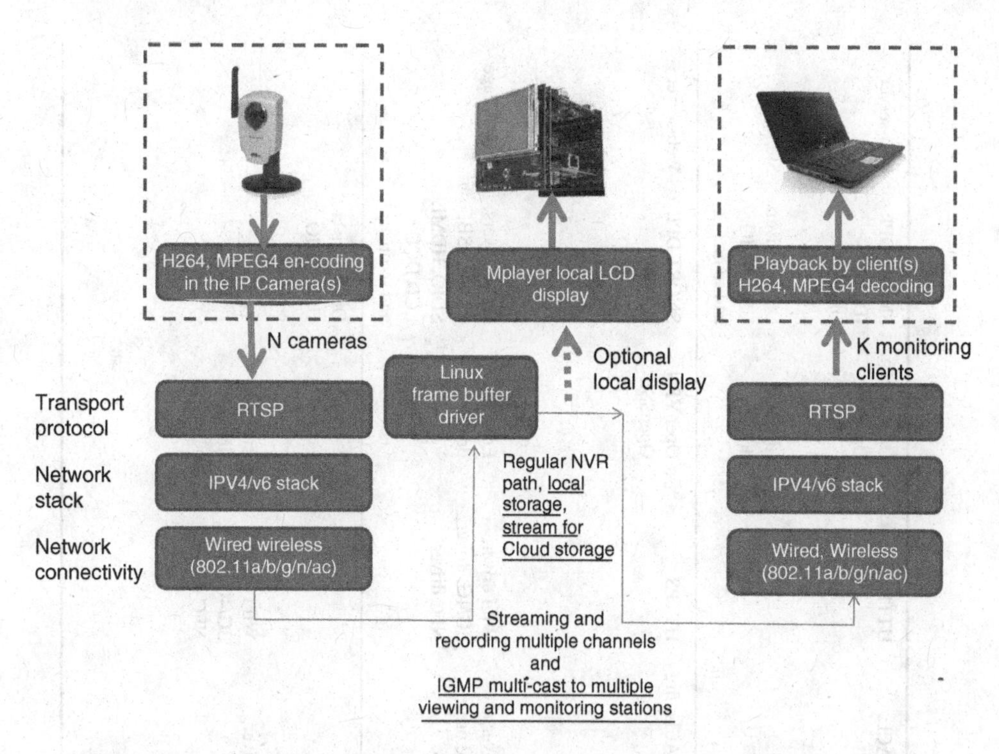

Figure 3.7 Video surveillance and network video recording (NVR) architecture

monitoring with facial recognition, license-plate recognition, automated assembly inspection, and so on.

Video Surveillance and Machine Vision at Industrial Market and Infrastructure Level

Video surveillance uses an NVR (Network Video Recorder) as shown in Figure 3.7. Video surveillance and NVR is a scalable approach to scale for implementing security at buildings, factories, and city level [8].

Table 3.3 shows software stacks for video surveillance and NVR.

Figure 3.8 illustrates video surveillance and NVR. The architecture, showing numerous IP cameras (~64 cameras), usually streams to a video server called the Network Video Recorder for video recording of camera streams and for easy communication with automation systems (SCADA-supervisory control and data acquisition). The IP cameras can send event tags to the automation system with system information and the status of each channel. In addition, the video server can receive event tags sent from the automation system to trigger video recording and other actions.

Table 3.3 Video surveillance and network video recording (NVR)

	CoAP	DB – SQL, unstructured DB	Security, IPSEC, SSL, DTLS, encryption	NAT, ACL, firewall	HTTPS	OpenVPN, OpenSSL	SNORT DPI, IDS/IPS – DDOS	Web-server
Service (cloud)	Networking ipv6, VLAN, DNS	Cloud apps – billing, and so on				Ethernet driver	Other drivers, USB, SDIO, and so on	Storage
	CoAP	DB – SQL, unstructured DB	Security, IPSEC, SSL, DTLS encryption, secured boot	NAT, ACL, fire-wall	HTTPS	OpenVPN, OpenSSL	SNORT DPI, IDS/IPS – DDOS	Web-server
IoT gateway (wireless router)	Networking, ipv6, VLAN, DNS, IGMP multi-cast, live 555 media server (NVR), RTSP	Provisioning and Mgt, TR69, QOS	Wireless security, WPA2, WEP	BT driver, ZigBee driver	WiFi driver, 3G/4G, driver, NFC driver	Ethernet driver	Other drivers, SATA, USB, SDIO, HDMI, LCD DCU, and so on	NAS storage
							Other drivers, USB, SDIO, and so on	
	ZigBee driver		Security, IPSEC, SSL, DTLS, encryption, secured boot	BT driver	WiFi driver, 3G/4G driver, NFC driver	Ethernet driver		
Sensing node			Wireless security, WPA2, WEP					

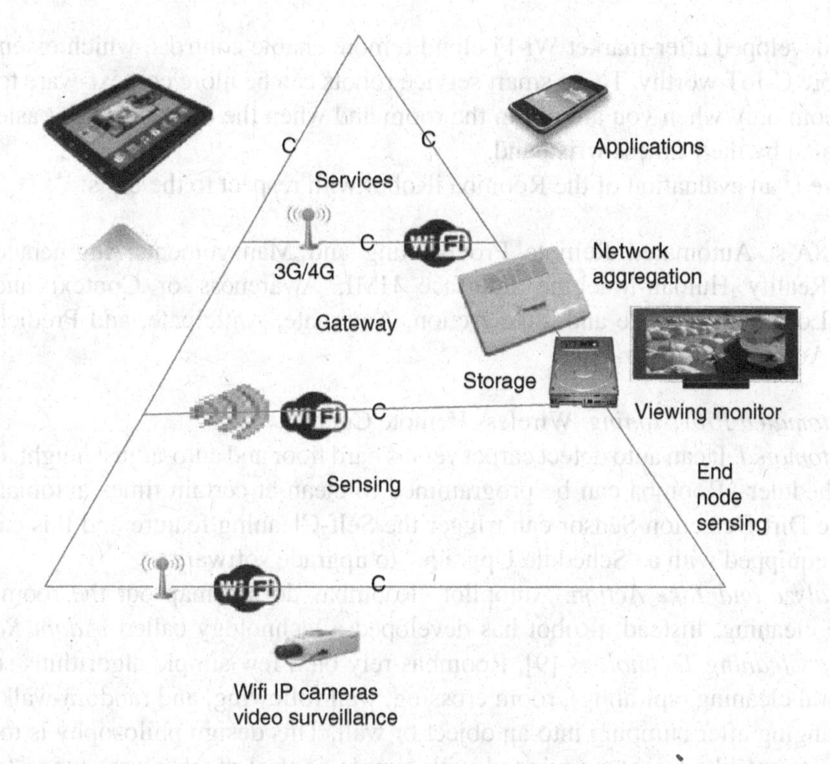

Figure 3.8 Video surveillance and network video recording (NVR)

Other features are as follows:

- remote live multi-cast viewing and remote playback via web access of transcoded videos with H.264, MPEG4, and MJPEG formats;
- intelligent key frame to decode only to save system resources for higher priority tasks;
- video recording with manual, event-triggered, and scheduled recording;
- playback system with event- and time-based search functionality;
- video analytics with face recognition, license-plate recognition, and so on.

3.1.3.4 Service Robots at Consumer Level – Roomba iRobot

Another Smart home device is the Roomba® vacuum cleaning robots from iRobot which also showcase integration of a rich set of smart software with sensors. Other home service robots for autonomous home automation can automate house chores such as mopping the floor, cleaning windows, pool cleaning, and mowing the lawn. These service robots are equipped with rich sets of sensors such as Vision Camera, Ultrasonic, Infrared (IR) sensor, Obstacle IR Sensor, Dirt Detection Sensor, Battery Status Sensor, Optical Floor Sensor, Gyro, and Accelerometer and integrated with the following sophisticated smart software. Roomba by itself is just an M2M automation device without cloud interface. However, the Roomba community and enthusiasts have developed after-market Wi-Fi cloud remote enable controls, which extend it to be more C-IoT worthy. These smart service robots can be more context-ware to clean the room only when you are not in the room and when the owners are not asleep, as indicated by their fitness wristband.

Here is an evaluation of the Roomba iRobot with respect to the 8A's:

> 8A's: Automated Remote Provisioning and Management, Augmented Reality Human-Machine Interface HMI, Awareness of Context and Location, Analyze and Take Action, Automate, Anticipate, and Predict, Autonomous, Attractive

- *Automated Provisioning.* Wireless Remote Control.
- *Automated.* It can auto detect carpet versus hard floor and auto-adjust height. iRobot Scheduler – Roomba can be programmed to clean at certain times automatically.. The Dirt Detection Sensor can trigger the Self-Cleaning feature and this can also be equipped with a "Schedule Upgrade" to upgrade software.
- *Analyze and Take Action.* Autopilot – Roombas do not map out the rooms they are cleaning. Instead, iRobot has developed a technology called *iAdapt Responsive Cleaning Technology* [9], Roombas rely on a few simple algorithms such as spiral cleaning (spiraling), room crossing, wall following, and random-walk angle changing after bumping into an object or wall. This design philosophy is to make robots act like insects, equipped with simple control mechanisms tuned to their environment. The result is that although Roombas are effective at cleaning rooms, they take several times as long as a person would to do the job. The Roomba may cover some areas many times and other areas only once or twice.
- *Aware of Context and Location.* Roombas can stay out of designated areas (Virtual Wall). The Battery Status Sensor can automatically trigger the Self-Charging (Homebase) capability by automatically getting it to return and dock.
- *Attractive.* It has an aesthetic look and feel.

Next-Generation Smart Service Robots

Future service robots will walk your dogs outside in the snow, water your plants when you are on vacation, wash your dishes and clothes when you are asleep, iron your shirts, and cook your meals. The next level of smart service robots will be cloud connected and could engage and interact with human beings such as talk and sing to you and your kids, as well as massage you with customizable near-human techniques. There is a high-growth service robots market for the use of PRs (personal robots) in homes and IR (Industrial Robots) in industry with improvements in motion planning, computer vision (especially scene recognition), natural language processing, and automated reasoning.

In industrial robot and drone markets for land, air, and water applications; there are Unmanned Ground Vehicles (UGVs) for precision farming or robotic sentry, Aerial robots are referred to as *Unmanned Aerial Vehicles* (*UAVs*) and underwater robots are called *autonomous underwater vehicles* (AUVs). These autonomous drones can also be used for delivery of goods, food, and strikes to areas where there could be risks.

3.1.3.5 Smart Home Gateway (Scale to Smart Building Automation)

A smart IoT Home Gateway is used when there are multiple end-node sensor devices (Smart Thermostat, smart garage door, smart LED lighting) with multiple users need to be continuously in operation even after the users have left the premises with their smart mobile phones.

This section describes a Smart Home Gateway based on an integrated open source applications platform [10]. This smart gateway is scalable to smart building automation for lighting control, smoke alarm detection, smart door, smart window shades, video surveillance, and so on.

Motivation

Figure 3.9 shows an IoT model for a Smart Connected Home and Building. The smart gateway provides a converged wireless platform supporting ZigBee-based

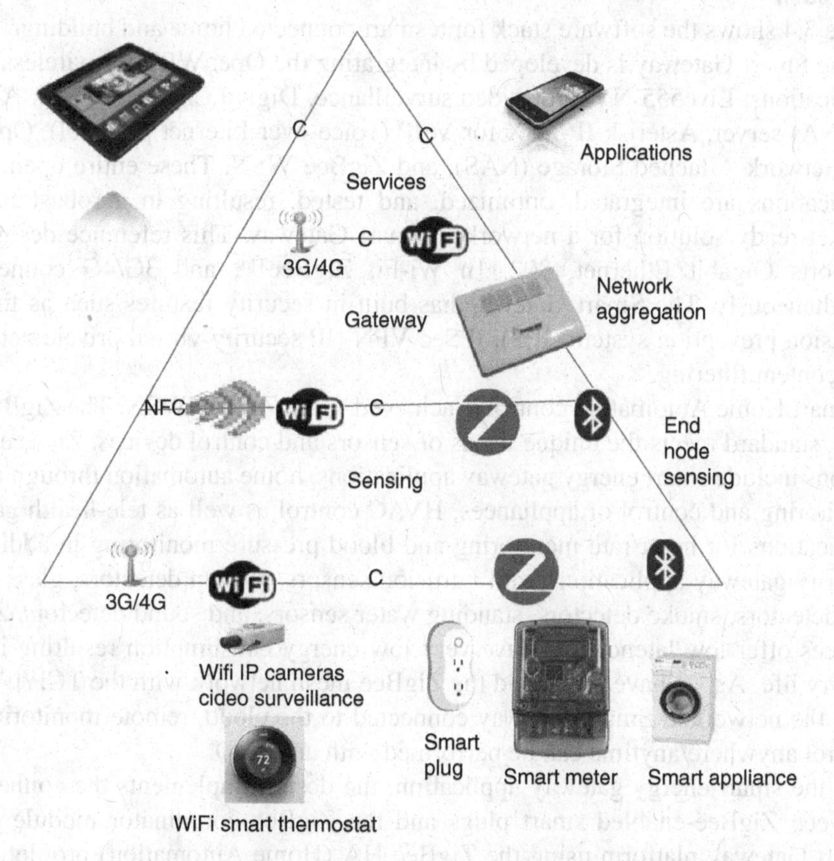

Figure 3.9 Smart connected home and building

WSN (802.15.4), WiFi (802.11), and 3G broadband connectivity integrated with Ethernet-based TCP/IP LAN and WAN network. Full sets of residential gateway services in safety, security, smart energy, and infotainment are supported. This smart gateway supports remote monitoring and control of smart metering and energy consuming appliances (residential/businesses) using Mobile Internet Devices (MIDs) such as smartphones, smart tablets from Apple, Android, or smart tablets connected to the Smart Gateway.

Regulation on power dissipation of home-based networking equipment is now common, and for good reason. There is more that can be done. Energy savings made as a result of effective home automation can be enhanced by bundling functionality and services onto a single M2M-enabled digital home gateway which can support IPTV, broadband wireless, media storage and distribution, medical and home automation, and more. A virtualized software platform can allow different service providers or utilities to run concurrently on the same box without interference. Visualization and control can be achieved by connecting via smartphone, TV, tablet, or netbook.

Approach

Table 3.4 shows the software stack for a smart connected home and building.

The Smart Gateway is developed by integrating the OpenWRT for wireless router applications, Live555 NVR for video surveillance, Digital Living Network Alliance (DLNA) server, Asterisk IP-PBX for VoIP (voice-over-Internet protocol), Openfiler for Network Attached Storage (NAS), and ZigBee WSN. These entire open source applications are integrated, optimized, and tested, resulting in a robust turnkey, market-ready solution for a networked Smart Gateway. This reference design also supports Gigabit Ethernet, 802.11n Wi-Fi, ZigBee™, and 3G/4G connectivity simultaneously. The Smart Gateway has built-in security features such as firewall, intrusion prevention system (IPS), IPSec VPN (IP security virtual private network), and content filtering.

Smart Home Automation control is achieved using ZigBee WSN. The ZigBee network standard meets the unique needs of sensors and control devices. ZigBee applications include smart energy gateway applications, home automation through remote monitoring and control of appliances, HVAC control as well as tele-health gateway applications for heart rate monitoring and blood pressure monitoring in addition to security gateway applications with intrusion sensors, motion detectors, glass breakage detectors, smoke detectors, standing water sensors, and sound detectors. ZigBee devices offer low latency and have very low energy consumption resulting in long battery life. As we have integrated the ZigBee mesh network with the TCP/IP stack, with the networked Smart Gateway connected to the cloud, remote monitoring and control anywhere/anytime can be performed with any MID.

In the smart energy gateway application, the design implements the connectivity between ZigBee-enabled smart plugs and the ZigBee coordinator module on the Smart Gateway platform using the ZigBee HA (Home Automation) profile. Appliances which have a built-in ZigBee module can be directly connected to the Smart

Table 3.4 Software stack for smart connected home and building

	CoAP	DB – SQL, unstructured DB	Security, IPSEC, SSL, DTLS, encryption	NAT, ACL, firewall	HTTPS	OpenVPN, OpenSSL	SNORT DPI, IDS/IPS – DDOS	Web-server
Service (cloud)	Networking, ipv6, VLAN, DNS	Cloud apps – billing, and so on				Ethernet driver	Other drivers, USB, SDIO, and so on	Storage
	CoAP	DB – SQL, unstructured DB	Security, IPSEC, SSL, DTLS, encryption, secured boot	NAT, ACL, firewall	HTTPS	OpenVPN, OpenSSL	SNORT DPI, IDS/IPS – DDOS	Web-server
IoT gateway (wireless Router)	Networking, ipv6, VLAN, DNS	Provisioning and Mgt, TR69, QOS	Wireless security, WPA2, WEP	BT driver, ZigBee driver	WiFi Driver, 3G/4G driver, NFC driver	Ethernet driver	Other drivers, USB, SDIO, and so on	Storage
							Other drivers, USB, SDIO, and so on	
	ZigBee driver		Security, IPSEC, SSL, DTLS, encryption, secured boot	BT driver	WiFi Driver, 3G/4G driver, NFC driver	Ethernet driver		
Sensing node			Wireless security, WPA2, WEP					

Gateway platform and the ZigBee-enabled Modlet enables traditional appliances also to communicate with the MPC8308 platform and be remotely monitored or controlled from anywhere at any time. This design enables IoT connectivity and M2M communication.

The connectivity between the ZigBee-enabled SE (Smart Energy) meter and built-in ZigBee module on the Smart Gateway platform is implemented using a ZigBee SE profile. In addition, several SE meters can be monitored via a Data Concentrator in the Neighborhood Area Network (NAN). The data concentrator can read the energy consumption data from each of the meters via Power Line Communication (PLC) connectivity and upload the data back to the utility server via 3G broadband. Utility companies can also push messages related to peak-load tariff rate change to each individual home through the smart meters.

Smart secure video surveillance is achieved though WiFi IP cameras supported by Live555 video media server running on the wireless gateway. The gateway also supports DLNA media streaming of videos and music that can be sent to multi-room, multi-users.

Results

- *Anywhere, anytime* remote monitoring and control of appliances using the Thinkeco smart plug based on ZigBee connected by a ZigBee mesh network hosted by the Smart Gateway using mobile devices through the cloud.
- Full Residential Gateway and infotainment functions
  - For example, video surveillance: Remote monitoring, recording and playback of video surveillance @36 Mbps using 12 cameras with D1 (3 Mbps) can be done simultaneously using NVR application. Higher-end multi-core processors can be used to scale the number of cameras to over 100 cameras (video data rate to be over 300 Mbps)
  - For example, high performance wireless DLNA media streaming and voice telephony based on high performance wireless access point (AP). An integrated 802.11n Wi-Fi module delivers over 300 Mbps of wireless local area network (WLAN) performance with bandwidth >120 Mbps. With high-end processor, 11ac WiFi can be supported that delivers >900 Mbps wireless throughput.

Conclusion

This Smart Gateway delivers a high-performance, integrated, optimized, and cost-effective solution with multiple applications running simultaneously.

Figure 3.10 shows Smart Connected Home Automation that provides Smart energy, safety, and security. The integration and optimization of multiple open source applications is well tested and validated resulting in a reliable and sustainable reference design solution. It enables remote monitoring and control *anywhere, anytime* using smart mobile devices for cloud services such as smart energy, Live555 NVR for video surveillance, Universal Plug, and Play (UPnP), DLNA, VoIP, 802.11n Wi-Fi, 3G broadband, and 802.15.4 ZigBee. High performance networking by bridging multiple wireless networks together into a wireless mesh network (ZigBee WSN, TCP/IP,

Figure 3.10 Smart connected home automation, smart energy, safety, and security

Wi-Fi, and 3G/4G) and high bandwidth wired network (1 Gbps Ethernet) enables multiple gateways for Smart energy, health, security, and residential gateway (ZigBee wireless sensors, Media, and Voice) of cloud application services.

3.1.3.6 Smart Building Automation

Buildings represent another area where energy efficiencies can be made. In offices, hotels, or campuses, it is reasonable to suggest that people do not have the same motivations as they do at home to conserve energy. The potential for M2M to automate is magnified for buildings because the goal is not only to save energy but also to implement security. CCTV video surveillance and secure access systems such as card swipe, card proximity, or something more advanced like iris scanning, have an increasing role to play in securing our towns and cities (Figure 3.11).

In addition to building automation such as LED lighting, AC temperature, and humidity control, sensors are also deployed for sensing structural issues of buildings and bridges, so that preventive care can be deployed before major collapse happens (Figure 3.12).

New buildings constructed with glass and steel as well as old stone constructions often suffer from poor in-building wireless coverage. A compelling option for in-building M2M networks could be a combination of wireless with wired

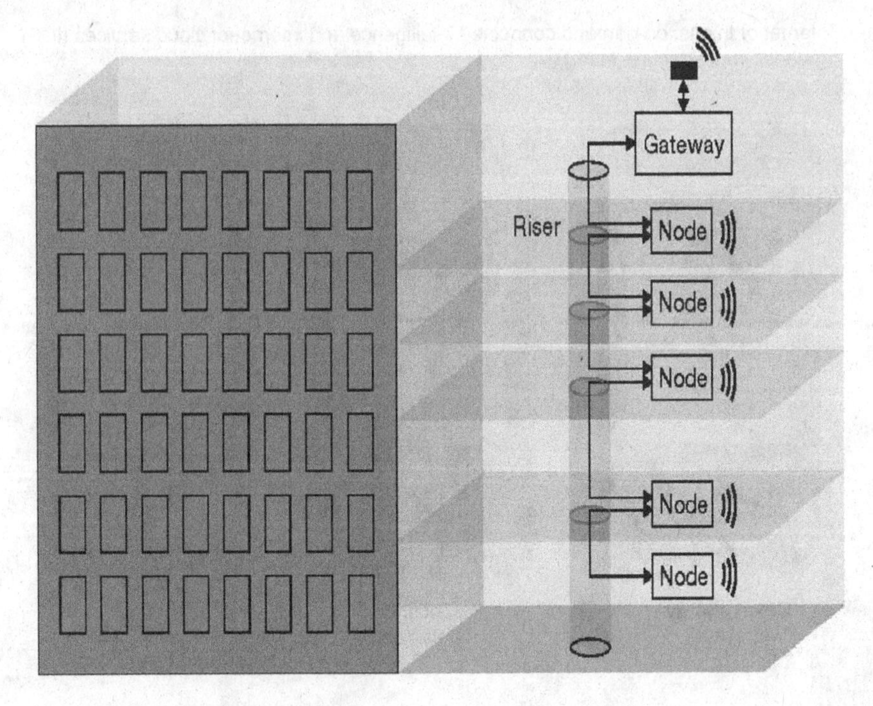

Figure 3.11 M2M network for buildings and factories

networking. Daisy-chained Ethernet, for example, can scale the height/length of a building with lower costs compared to the more traditional star or switched network. Wireless can extend network reach across individual rooms or floors.

3.1.4 Smart Energy and Smart Grid

This section describes Smart Grid applications in all three domains (consumer home, industrial, and infrastructure) as shown in Figure 3.13.

This includes smart meters and smart data concentrators that are used for Advanced Metering Infrastructure (AMI) that also provides the essential Demand Response (DR) for preventing brownouts. Figure 3.14 shows Smart Energy with Smart meter, home area network (HAN), NAN.

Figure 3.16 shows Potential Energy Saving with Smart Energy Management.

3.1.4.1 Introduction

Local networking of electronic devices in houses and buildings offer benefits in a number of areas, ranging from safety and security to energy efficiency and home entertainment features. HANs can be implemented via both wired and wireless solutions, using multiple different standards, and can be remotely controlled and monitored

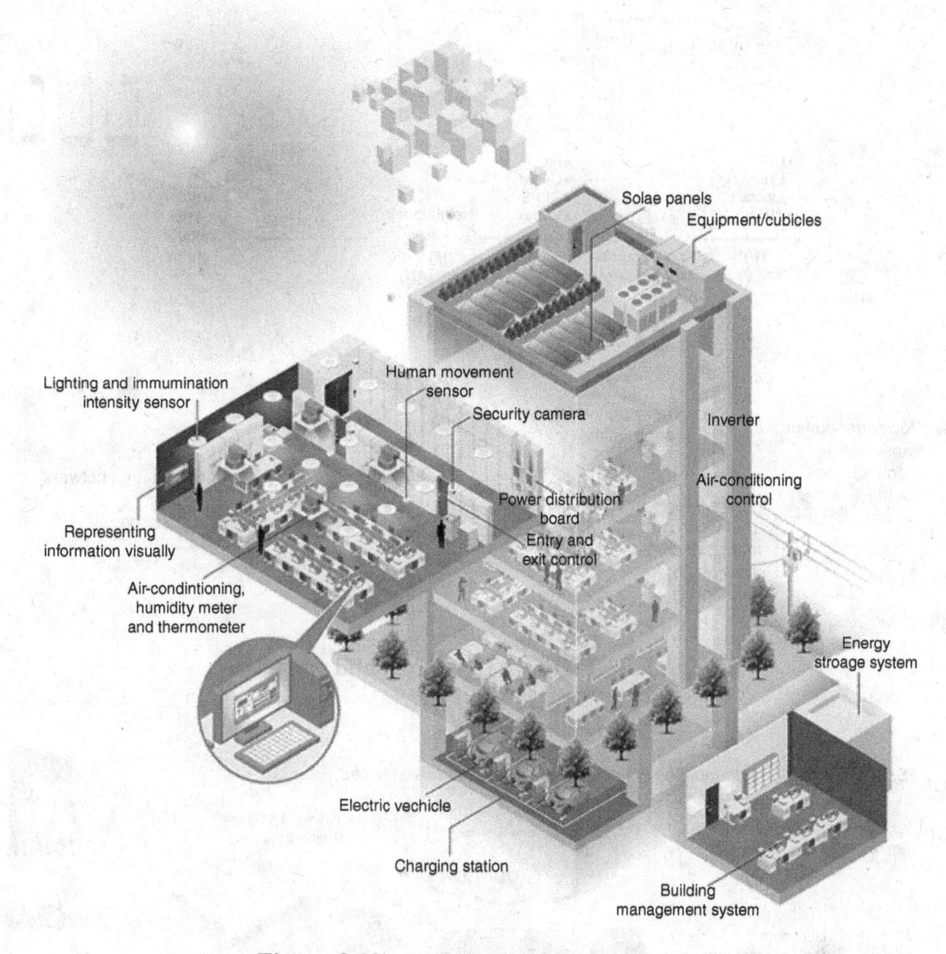

Figure 3.12 M2M industrial automation

through a gateway to neighbor, wide area, or smart grid networks. Figure 3.15 shows a model of Smart Energy (Smart Grid and Metering) [21]. Table 3.5 shows Software Stacks for Smart Grid and Metering.

While smart grid deployments offer great opportunities for utilities to manage and control energy distribution to their customers, it also gives homeowners the opportunity to better manage their energy usage through smart energy management (Figure 3.16).

3.1.4.2 Home Area Networks (HANs)

A HAN is a dedicated network connecting devices in the home such as displays, load control devices, and ultimately "smart appliances" seamlessly into the overall smart

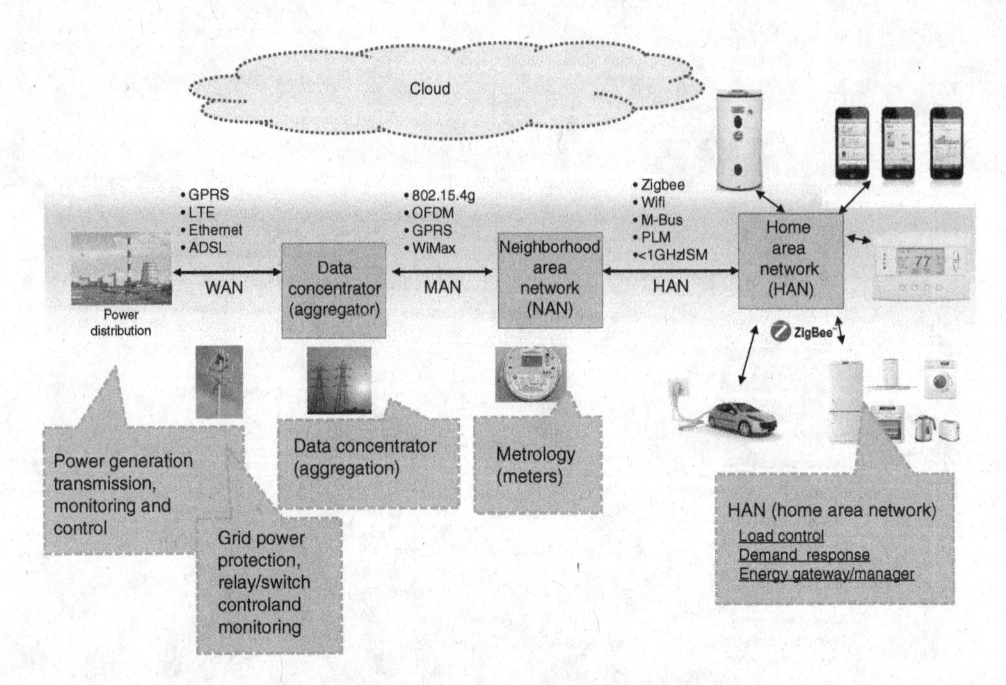

Cloud
• GPRS
• LTE
• Ethernet
• ADSL
WAN
• 802.15.4g
• OFDM
• GPRS
• WiMax
MAN
• Zigbee
• Wifi
• M-Bus
• PLM
• <1GHzISM
HAN
Power distribution
Data concentrator (aggregator)
Neighborhood area network (NAN)
Home area network (HAN)
ZigBee
Power generation transmission, monitoring and control
Grid power protection, relay/switch controland monitoring
Data concentrator (aggregation)
Metrology (meters)
HAN (home area network)
Load control
Demand response
Energy gateway/manager

Figure 3.13 Smart grid

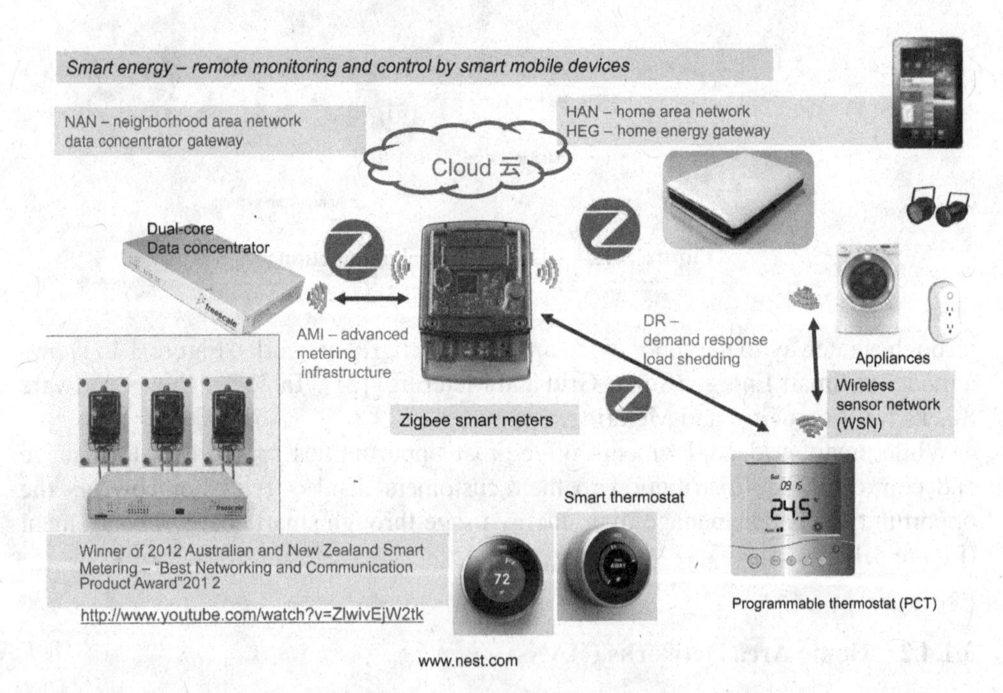

Smart energy – remote monitoring and control by smart mobile devices
NAN – neighborhood area network
data concentrator gateway
HAN – home area network
HEG – home energy gateway
Cloud 云
Dual-core
Data concentrator
AMI – advanced metering infrastructure
Zigbee smart meters
DR –
demand response
load shedding
Appliances
Wireless sensor network (WSN)
Smart thermostat
Winner of 2012 Australian and New Zealand Smart Metering – "Best Networking and Communication Product Award"201 2
http://www.youtube.com/watch?v=ZlwivEjW2tk
www.nest.com
Programmable thermostat (PCT)

Figure 3.14 Smart grid with smart meter, HAN, and NAN

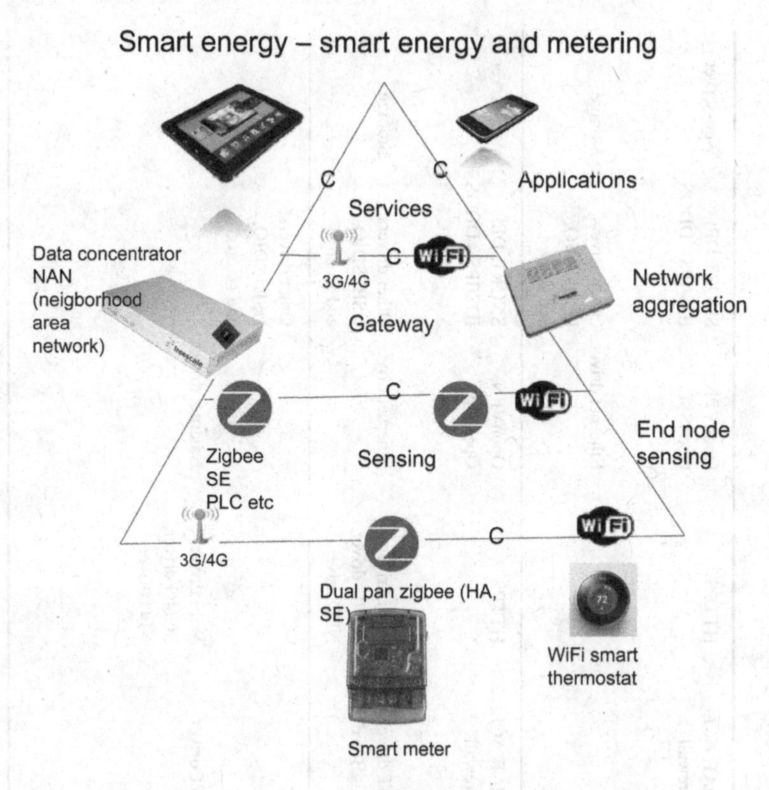

Figure 3.15 Smart energy (smart grid and metering)

metering system. It also contains turnkey reference designs of systems to monitor and control these networks. Most of our high-energy use today comes from heating/cooling, cooking, lighting, washing, and drying. These home appliances are beginning to become smart with connectivity features that allow them to be automated in order to reap benefits that smart metering and variable tariffs bring. The utility companies are beginning to be able to better manage the energy demand and perform load balancing more efficiently.

Realizing long-term potential savings in a typical home environment through the smart grid means that technology, legislation, and mind-set must come together to drive a permanent change in the way that consumers perceive energy consumption. Figure 3.17 shows a HAN example.

3.1.4.3 Strong HAN Market Growth

According to IMS and Pike Research, the installed base of smart home networks (majority are equipped with home energy management) will increase 4× from 14.7 million homes in 2014 to ~60 million in 2020.

Table 3.5 Software stacks for smart energy

	CoAP	DB – SQL, unstructured DB	Security, IPSEC, SSL, DTLS, encryption	NAT, ACL, firewall	HTTPS	OpenVPN, OpenSSL	SNORT DPI, IDS/IPS – DDOS	Web-server
Service (cloud)	Networking, ipv6, VLAN, DNS	Cloud apps – billing, and so on				Ethernet driver	Other drivers, USB, SDIO, and so on	Storage
	CoAP	DB – SQL, unstructured DB	Security, IPSEC, SSL, DTLS, encryption, secured boot	NAT, ACL, firewall	HTTPS	OpenVPN, OpenSSL	SNORT, DPI, IDS/IPS – DDOS	Web-server
IoT gateway (wireless router)	Networking, ipv6, VLAN, DNS	Provisioning and Mgt, TR69, QOS	Wireless security, WPA2, WEP	BT driver, ZigBee driver	WiFi driver, 3G/4G driver, NFC driver	Ethernet driver	Other drivers, USB, SDIO, and so on	Storage
							Other drivers, USB, SDIO, and so on	
	ZigBee driver		Security, IPSEC, SSL, DTLS, encryption, secured boot	BT driver	WiFi driver, 3G/4G driver, NFC driver	Ethernet driver		
Sensing node			Wireless security, WPA2, WEP					

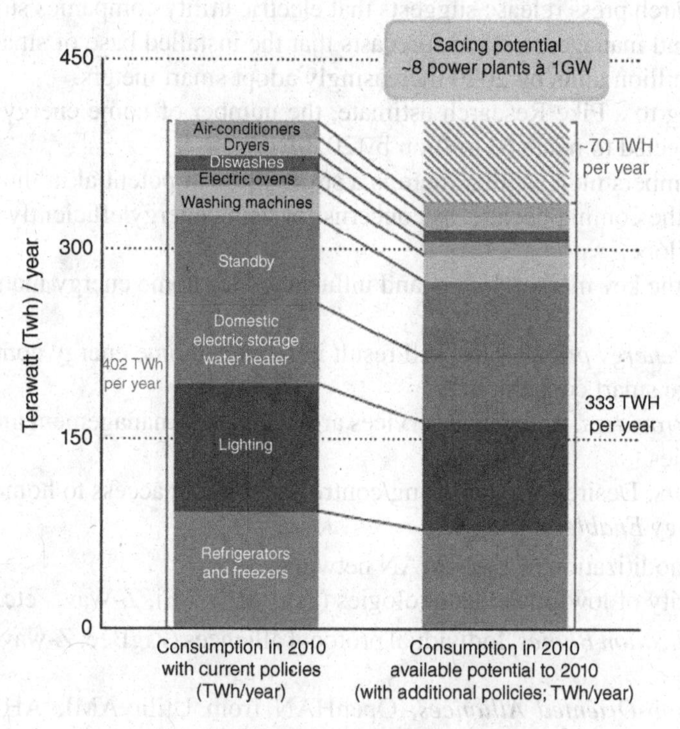

Figure 3.16 Motivation on smart energy management

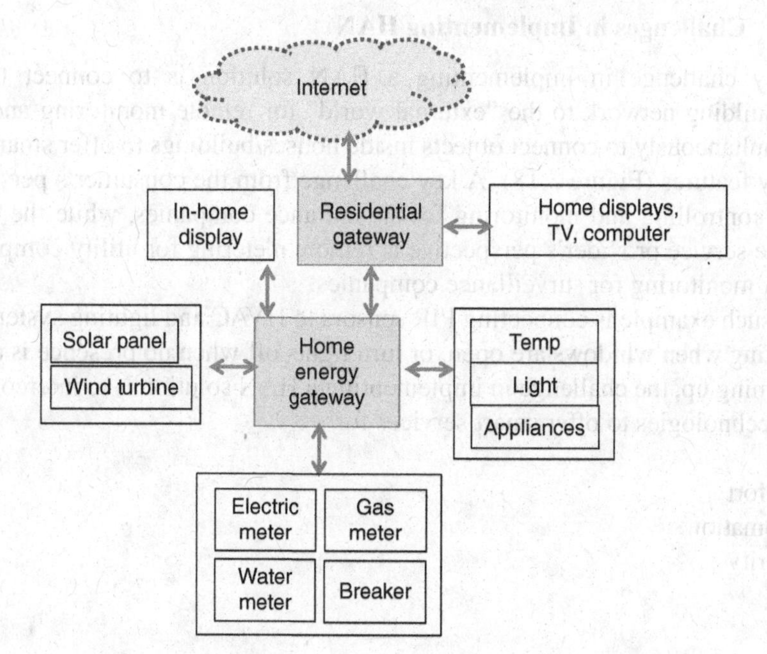

Figure 3.17 Home area network (HAN)

Pike research press release suggests that electric utility companies supports energy efficiency and management and forecasts that the installed base of smart meters will reach 963 million units by 2020 increasingly adopt smart meters.

According to a Pike Research estimate, the number of home energy management users is expected to reach 63 million by 2020.

These numbers indicate that there is a strong growth potential in the HAN market at least for the coming decade, as concerns for using energy efficiently are spreading across the globe.

Some of the key market drivers and influencers for home energy management are

- *Growing energy prices*. This will result in reduced home energy consumption and encourage smart consumption
- *Service Providers*. Innovative services around energy management and home security utilities.
- *Consumers*. Desire for monitoring/controlling remote access to home
- *Technology Enablers*:
  - Commoditization of LAN/WAN networks
  - Maturity of low-power technologies (ZigBee, Wi-Fi, Z-Wave, etc.)
- *Standardization Bodies*. Individual protocol alliances (ZigBee, Z-wave, HomePlug, etc.)
- *Application-Oriented Alliances*. OpenHAN from UtilityAMI, AHEM, CECED from appliances.

3.1.4.4 Challenges in Implementing HAN

The key challenge in implementing a HAN solution is to connect the entire house/building network to the "external world" for remote monitoring and control, and simultaneously to connect objects inside houses/buildings to offer smart interoperability features (Figure 3.18). A key challenge from the consumer's perspective is remote controlling and monitoring for surveillance companies, while the challenge from the service provider's perspective is remote metering for utility companies and security monitoring for surveillance companies.

One such example is connecting PIR sensors to HVAC and lighting systems to turn off heating when windows are open, or turn lights off when no presence is detected.

Summing up, the challenge in implementing a HAN solution is to interconnect different technologies to offer smart services for

- Comfort
- Automation
- Security

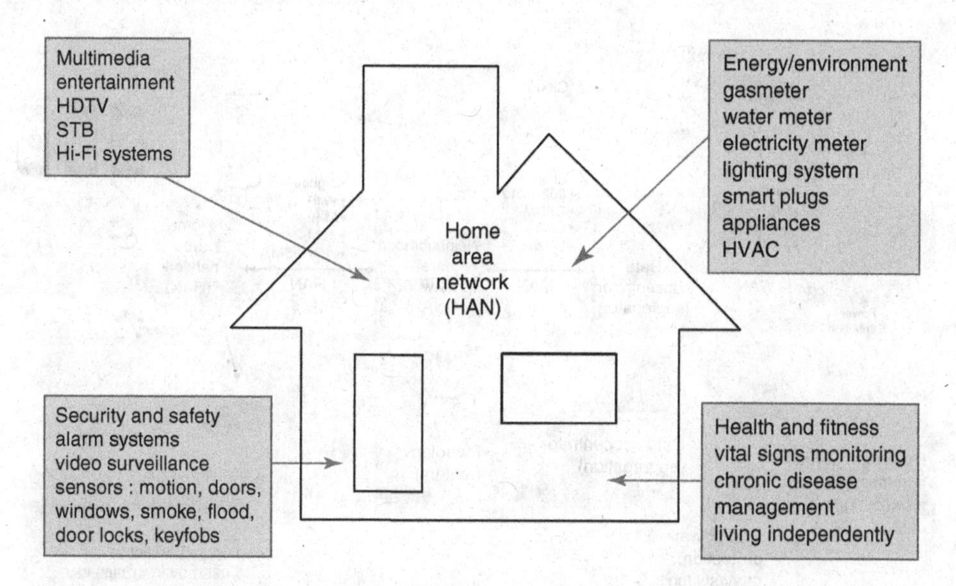

Figure 3.18 Smart networked home forecast

- Energy management
- Health.

3.1.4.5 Smart Energy Solution

A smart energy solution consists of a HAN with smart appliances and electric meters at one end and grid-end applications such as data concentrator/aggregator, grid routers, and grid power management and protection at the other end (Figure 3.19). The ZigBee smart energy application profile addresses communication from meter to the HAN for purposes of load control and DR. Load control provides the ability for the utility to turn off loads for short periods of time in the customer's premises during peak loads, while DR is the ability for utilities to communicate with a home the changing utility rates during peak times and similar details. The user will then have an option of taking voluntary action to reduce personal consumption.

3.1.4.6 Smart Energy Gateway

An energy gateway is the interface between the utility-controlled smart grid and energy-consuming in-house objects. Most utility providers prevent direct access to smart meters. The utility providers transmit the smart meter readings to utility servers via the data concentrators. Then, consumers need to connect to the utility server to have access to meter readings. It would be more likely for smart energy gateway to access the main fuse box or the Smart Thermostat.

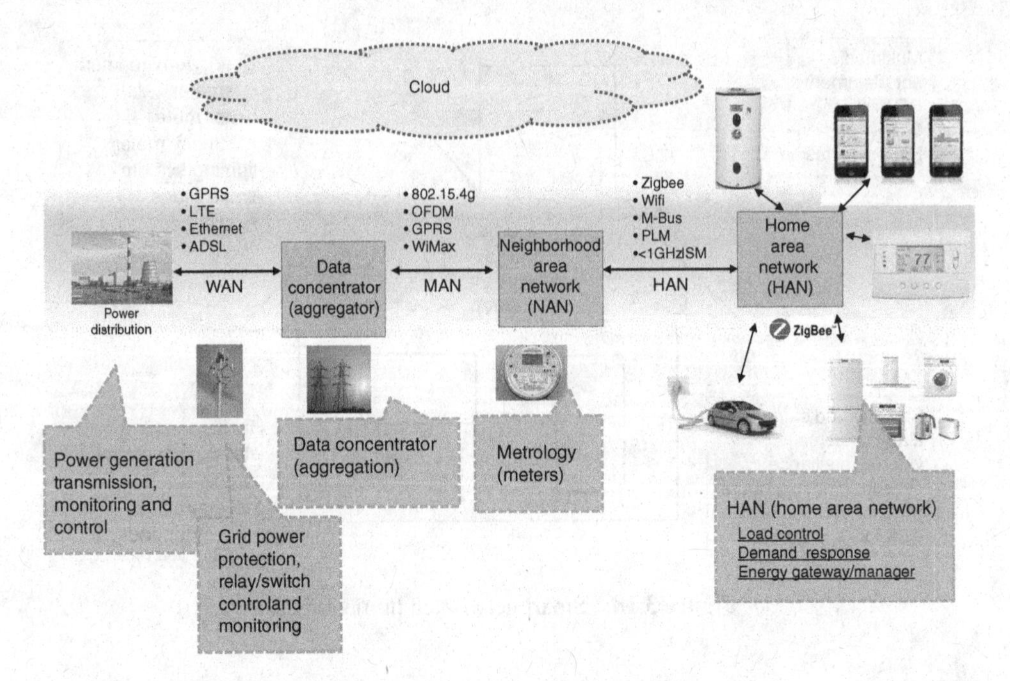

Figure 3.19 Smart energy solution

3.1.4.7 Data Aggregators/Concentrators in Neighborhood Area Network (NAN)

A data concentrator (Figure 3.20) is an important component in automatic meter reading (AMR) [11].[1] More importantly, data concentrators are essential in AMI that provides Demand Response (DR) in load-balance to avoid brownout. It creates the necessary network infrastructure by linking several utility meters (electricity, gas, water, heat) to the central utility server and captures and reports vital data. It also helps synchronize the time and date data of utility meters to a central utility server and enables secure data transfer of user authentication and encryption information. Communication to utility meters is comprised of an RF or wired (power line modem) connection, enabling data transfers to the central utility server via GPRS, Ethernet, and GSM, POTS, or UHF/VHF networks. A data concentrator usually supports the device language message specification (DLMS)/COSEM client/server stack standard to work with multiple meter-vendors.

Figure 3.21 shows a block diagram for an implementation of a data concentrator or aggregator and grid router.

The key functions of the data concentrator are as follows:

- AMI – Energy data collection/aggregation and DR management
- Secure data routing
- Packet time stamping.

[1] Wu was awarded 2012 Innovator of the Year by ECD for his platform approach to Wireless Smart Gateways.

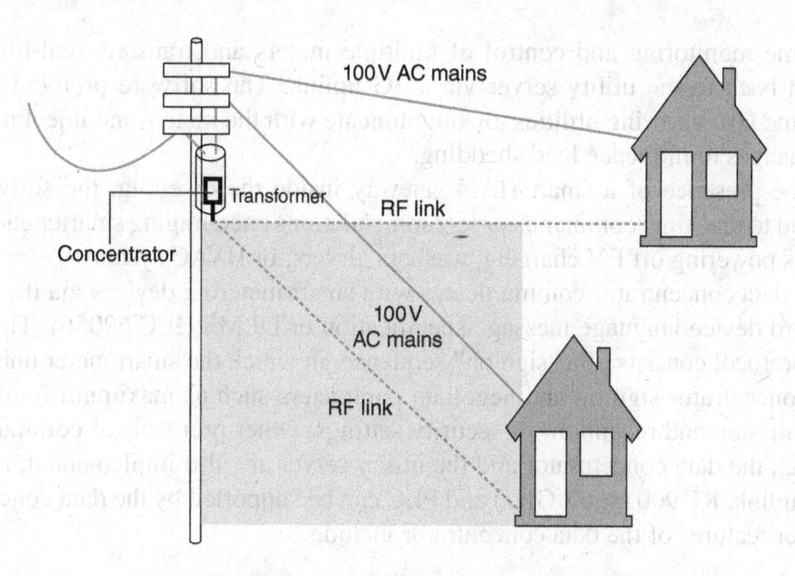

Figure 3.20 Typical data concentrator setup

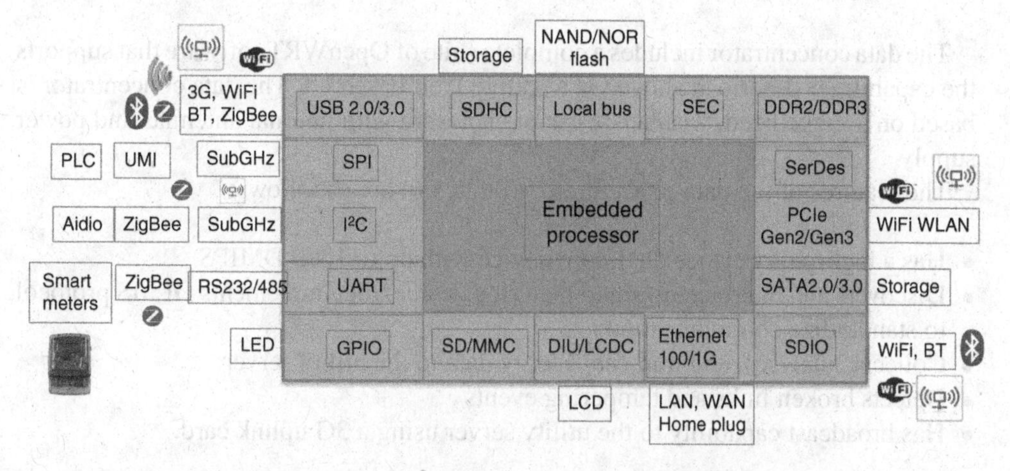

Figure 3.21 Data concentrator and grid router block diagram

3.1.4.8 Data Concentrator

The data concentrator reference design enables communication to smart grid meters within an NAN. The reference design is capable of a variety of usage models, including smart energy device discovery, communication protocols, and uplink communication to the utility server. The embedded processor used can deliver dual-core performance running up to 533 MHz, enabling a variety of complex usage models within a low power envelope that allows for fanless designs.

The data concentrator will instantly discover and connect with multiple smart energy meters. Enabled with an AMI, the data concentrator allows for bidirectional real-time monitoring and control of multiple meters and transfers real-time information back to the utility server via a 3G uplink. The software protocol supports real-time DR, allowing utilities to communicate with the meters and inject peak-load rate changes to influence load shedding.

In the presence of a smart HAN gateway inside the dwelling, the software can respond to real-time commands or recommendations encouraging smarter energy use, such as powering off EV chargers, washers, dryers, or HVAC systems.

The data concentrator communicates with smart metering devices via the industry standard device language message specification or DLMS (IEC 62056). The widely used protocol consists of a "sign on" sequence, in which the smart meter unit and the data concentrator sign on and negotiate parameters such as maximum frame length (transmission and reception) or security settings. Other protocols of communication between the data concentrator and the utility server are also implemented, including a 3G uplink. RF 900 (sub 1 GHz) and PLC can be supported by the data concentrator.

Other features of the data concentrator include

1. Detection and reporting of line breaks to the utility company
2. Alerting the utility company of smart meter tampering.

The data concentrator includes a complete suite of OpenWRT software that supports the capabilities described above via a simple Web-based UI. The data concentrator is based on a ruggedized, weather-resistant enclosure with internal antennas and power supply.

The features of the data-concentrator (Table 3.6) are as follows:

- Has a high-performance dual-core device with up to 1300 DMIPS
- Discovers and interfaces to smart metering devices and implements DLMS protocol to standardize communication
- Collects, analyses, and transfers energy data to the utility server
- Detects broken links and tampering events
- Has broadcast capability to the utility server using a 3G uplink card.

3.1.4.9 Grid Router

The grid router's main function (Figure 3.22) is to provide secured connectivity interface between the smart meter and the utility network, performed using a grid router (sometimes referred to as a *concentrator*). The role of the router is to provide a link from the utility company to all local smart meters, usually running a real-time operating system and provide high-level services such as communications stacks, message prioritization, store/forwarding, network routing, and discovery.

Table 3.6 Data concentrator features

Processor	Processor 667 MHz dual core device Capabilities for IEEE1588 time stamping and security acceleration
Connectivity	Serial line drivers for communication to power line communication controllers 3G, WiMax, or WCDMA communication via USB interface Three Giga-bit Ethernet ports to enable WAN/LAN communications with ipv4, ipv6 Time stamping via IEEE 1588 protocol Encryption capability leveraging the device's IPSEC security accelerator Supports DLMS IEC 62056 protocol
Memory	Up to 128 MB of NOR/NAND flash memory Capability to interface to DDR2/3 memory up to 800 MHz data rate
Enclosure and design	Energy efficient passive cooled design, natural convection capable Ruggedized, weather resistant construction
Future development	Power line communication (PLC) and sub-gigahertz RF interfacing and protocol development

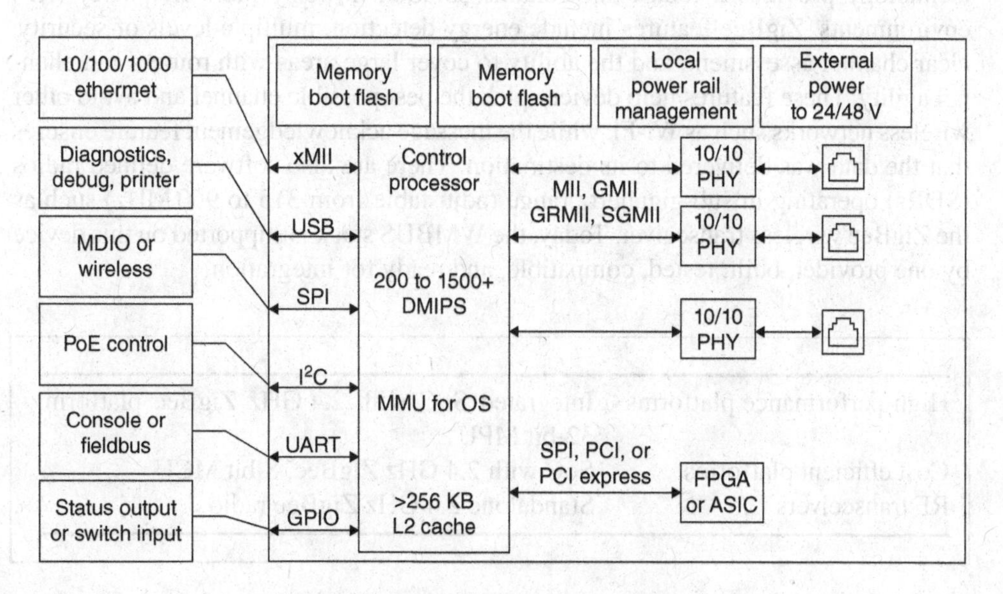

Figure 3.22 Grid router block diagram

Below are some of the key features that distinguish grid router solution:

- High performance (100 up to 38 000 MIPs)
- Built-in security functions supporting public and private key cryptography
- Wide range of communication ports, including Gigabit Ethernet and fast serial ports, plus USB 2.0 for local on-board interfacing
- Secured Connectivity.

3.1.4.10 Secured Connectivity

Depending on local needs, various options for wireless communication include short range wireless (sub-gigahertz) through 2.4 GHz ZigBee alliance and PLC (low frequency carriers typically below 500 kHz) using power line modem solutions for local communications and options from longer range communications such as ZigBee, Wi-Fi, Ethernet, ISDN, HDMI, PLC, Bluetooth/BLE, RF4CE, HomePlug, Z-Wave, and GPRS through strong alliance with leading smart grid standards bodies and committees.

3.1.4.11 ZigBee and Smart Energy

ZigBee (Figure 3.23) is a low-power wireless communications technology designed for monitoring and control of devices. Based on IEEE 802.15.4 standard, ZigBee technology provides a robust and reliable solution in noisy radio frequency (RF) environments. ZigBee features include energy detection, multiple levels of security, clear channel assessment, and the ability to cover large areas with routers and channel agility. These features help devices pick the best possible channel and avoid other wireless networks such as Wi-Fi, while the message acknowledgement feature ensures that the data was delivered to its destination. There are also software defined radios (SDRs) operating in sub-gigahertz range (adjustable from 315 to 960 MHz) such as the ZigBee wireless transceiver. Today, the WMBUS stack is supported on this device by one provider, built, tested, compatible, and ready for integration.

High-performance platforms	Integrated SoC with 2.4 GHz ZigBee platform, 32-bit MPU
Cost efficient platforms	SoC with 2.4 GHz ZigBee, 8-bit MCU
RF transceivers	Standalone 2.4 GHz ZigBee radio

Figure 3.23 ZigBee® and smart energy

3.1.4.12 Security

With so many forms of communication, security of these communications and that of the grid is paramount. The embedded processor selected should support an expansive range of security protocols and functions for both private and public key cryptography to help ensure that these links are protected from external attacks. For low data rate communication, AES and DES are commonly used. Since these are private key cryptography functions, extra care is needed to help ensure system integrity. The processor should also provide secure on-board storage of the keys to provide enhanced security for local communication.

3.1.4.13 Grid Power Management and Protection

Electricity substations are under increasing pressure to provide functionality to actively manage the local grid. Deployment of new, high performance power controller systems is now common across all regions as utility companies attempt to stay one step ahead of the technical challenges they face. Power relay controllers must manage surges and loading on the grid locally. Real-time control is essential to maintain grid integrity.

This is accomplished using embedded processors that support the IEEE 1588 protocol. A move toward cost-effective functional integration brings the focus toward multicore MPUs. Today, dual core is adequate for providing a cost-effective, comprehensive range of fast serial communications and dual-core performance.

Adding power meter functions in the power breaker provides more information on grid performance and loading. MPUs run SCADA for substation control and must have fast response times for the management of transient events (such as surges).

3.1.4.14 Power Efficiency

Low power operation is also an important consideration for MPU choice. Fanless operation greatly improves overall system reliability as these units may be installed in remote substations and they have a long maintenance cycle. Power Architecture products are highly efficient and many are designed for fanless operation, delivering gigahertz class performance at below 3 W.

3.1.4.15 Conclusion

The need for more efficient use of energy has led to the growth of the smart grid. Companies and government are enabling this management through the deployment of devices designed to efficiently manage power in the home and on the grid network.

3.1.5 Smart Energy Gateways

At the most basic level, there are two primary initiatives underlying the smart grid: generating clean, sustainable power and intelligently managing the distribution and use of that power. Achieving the transition to clean power generation will take many years, with an accompanying investment of billions of dollars. On the other hand, the means to intelligently manage the distribution of power and reduce how much each of us consumes is not only possible today, it is essential if we need to satisfy the escalating global demand for energy until new, clean, power generation sources come online. In order for utility companies to intelligently balance and distribute power, they need the ability to see where, how much, and when energy is being consumed. To create this ability, a fundamental feature of the smart grid is enabling the power distribution network to support the bidirectional flow of both power and communication capabilities from power distribution facilities to consumption locations. In more detail, this two-way communication extends from distribution centers out to urban clusters or neighborhoods and then branches out to individual residences and businesses which are connected to the smart grid by their electric meters, and increasingly, to individual devices within the premises itself. This network grid is illustrated in Figure 3.19. Just as individual computers, printers, storage systems, and servers are connected over Ethernet or wireless Ethernet (Wi-Fi) networks, the smart grid will connect devices that transmit, monitor, and consume electricity using a variety of new smart grid networking standards, including PLC, M-Bus, ZigBee technology, and Smart Energy 1.0 [12].

The smart grid device that enables utility companies to capture customer usage data is the smart meter, represented in Figure 3.24 by the metrology symbol in the middle of the diagram. The evolution of smart meters has been incremental, beginning with the integration of short-range RF technologies that allowed "drive-by" capture of meter readings, which saved time and improved accuracy. Next has followed smart meter improvements that have completely eliminated the need for mobile field staff to capture meter data. This latest round of smart meter improvements has been based on standards-based communication technologies driven by the AMI organization. The use of standards-based communication technologies has opened the door for existing home networking platforms, such as residential gateways and broadband AP routers, to incorporate support for them as well. This new class of platform is referred to as a "smart energy gateway," or alternatively, a "home energy gateway," and represents the heart of the HAN.

The key modifications necessary to enable a residential gateway to serve as a smart energy gateway is support for the physical layers (PHYs) and protocols that have been adopted for use in smart meters for communication. The PHYs used in smart meters include PLC, ZigBee, and 802.11. The associated communication protocols include DLMS, Smart Energy 1.0, and M-Bus in the European market. The integration of these PHYs, together with support for the communication protocols associated with them, are key requirements for designing a smart energy gateway. Increasingly, these

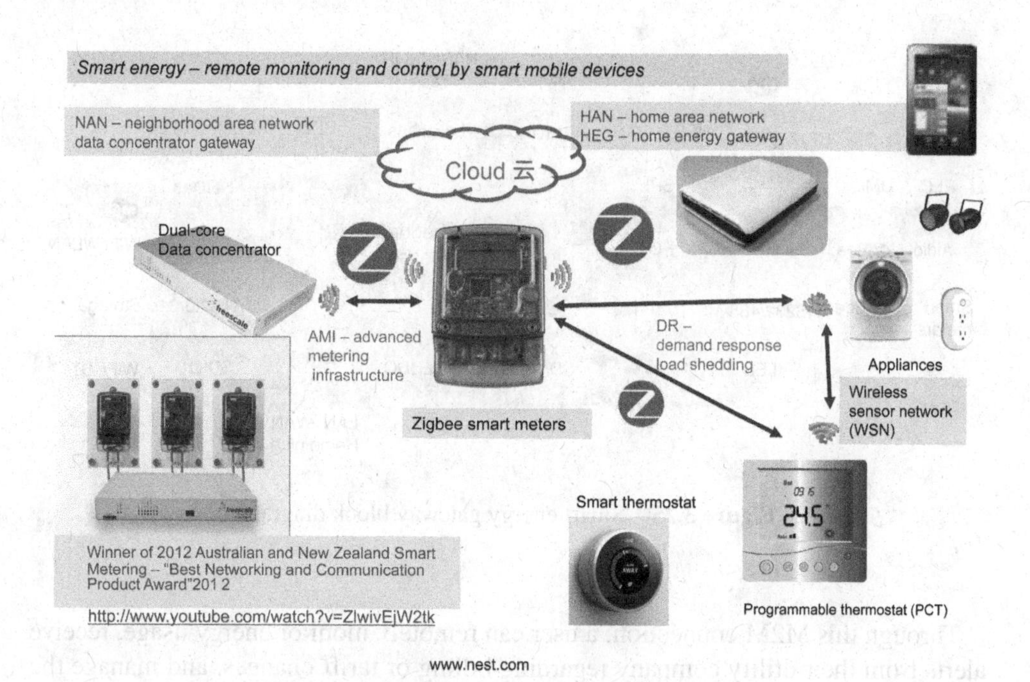

Figure 3.24 Home area network

same PHY technologies are rapidly being adopted for use within the HAN to support home automation connectivity with appliances, lighting, security systems, and health monitoring devices, as illustrated in Figure 3.25. Both the AMI organization as well as the Association of Home Appliance Manufacturers (AHAM) is guiding this enablement. Other considerations that must be factored into smart energy gateway designs include support for WAN access, and perhaps most importantly, a UI that enables access, monitoring, and control over the connected HAN devices.

This last factor deserves special emphasis. The ability to access, monitor, and control devices within the HAN is an essential capability if consumers are going to successfully manage and conserve the energy they use. The UI must allow customers to see exactly how much energy they are using, how much the utility company is charging for that energy, and provide the ability to exercise control over HAN-connected devices if necessary. Of equal importance to providing this UI is the ability to access it remotely, at any time, via a smart handheld device or tablet. This capability leverages the global preference by consumers to utilize a single smart device for all their communication applications, extending from voice, texting, email, and entertainment, to now include home monitoring, security, and control.

One example of these efforts is a networked Smart Energy Gateway (nSEG) reference design, shown in Figure 3.25. This multifunctional gateway can support M2M connectivity from smart handheld devices, such as smartphones or tablets, to the HAN.

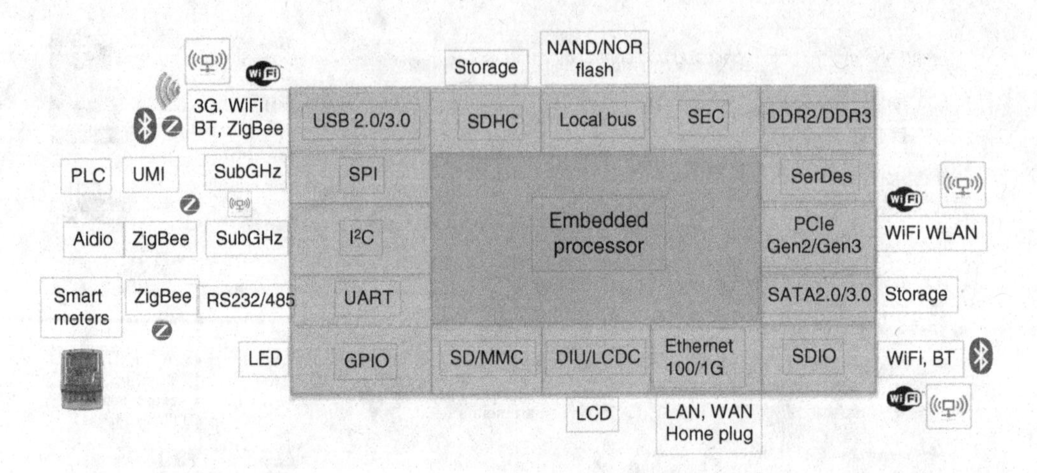

Figure 3.25 Smart energy gateway block diagram

Through this M2M connection, a user can remotely monitor energy usage, receive alerts from their utility company regarding billing or tariff changes, and manage the smart devices within their HAN.

The smart gateway enables connectivity between a variety of ZigBee-enabled HAN devices (with provision for powerline modem support) allowing them to be accessed and controlled over an M2M link from any smart handheld device via its GUI, as shown in Figure 3.26. In addition, the nSEG supports the latest 3 × 3 802.11n Wi-Fi radio modules via its PCI Express port, together with high performance gigabit Ethernet to enable true broadband connectivity. For wireless broadband applications, the 3G or 4G USB module, are supported via the two high-speed USB2.0 ports. The nSEG reference design kit includes a comprehensive suite of license-free OSGI software, including gateway, NVR, NAS, and DLMS stacks.

The low power home energy gateway reference platform features are as follows:

- A powerful, low power consumption applications processor that integrates a power management unit, a cryptography unit, and a rich set of connectivity controllers
- Dual ZigBee radios (with provision for power line modem) to enable seamless, plug and play connectivity to smart meters and the HAN automation system
- A WLAN wireless radio or Ethernet wire line interface (with provision for 3G/GPRS modem) to enable secure end-to-end HAN control and monitoring, either online or remotely, through a broadband access to the Internet
- A display interface to enable household management through an engaging and intuitive user interface.

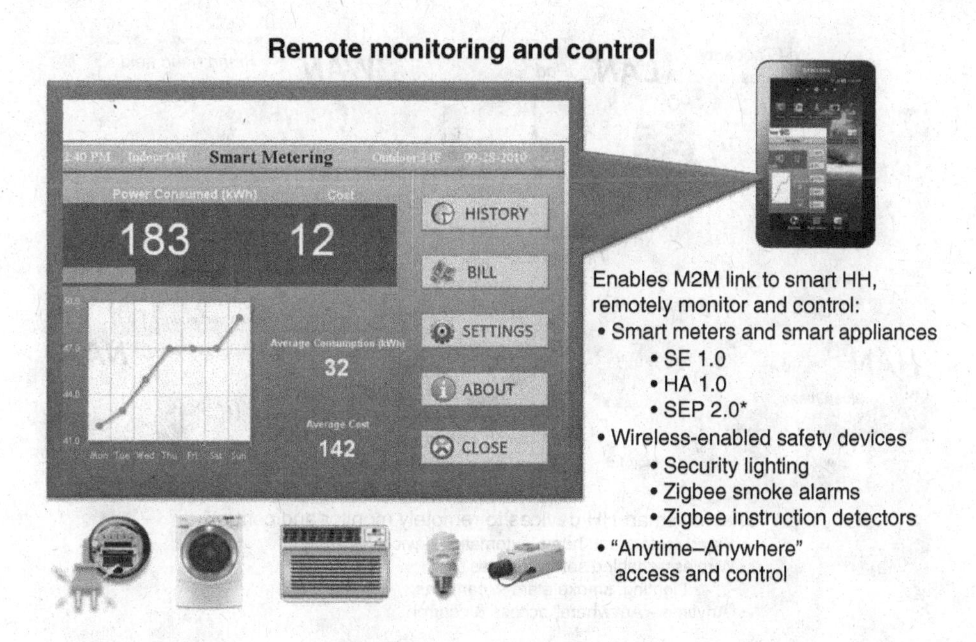

Figure 3.26 Graphical user interface to interact with smart gateway

3.1.5.1 Overview of Smart Gateways for Energy Management

Governments worldwide are mandating improved energy efficiency, requiring an investment in the new smart grid and smart energy management structure. The goal is to create a smart grid that will change the way power is deployed for sustainable energy around the world. At the heart of the worldwide rollout of smart meters and the construction of a smart grid network infrastructure lies the goal of energy efficiency from the generation, transmission, and distribution to the end customer. Leveraging the deployment of communications-enabled electricity meters, many applications can be offered to homeowners for optimizing overall energy management and to utility companies as a means of managing the load of their grid and preventing power demand peaks. Energy gateways are the interface between the utility-controlled smart grid and energy consuming in-house objects.

3.1.5.2 Networked Smart Gateways (NSG)

The smart grid device that enables utility companies to capture customer usage data is the smart meter, as a part of the HAN on the left side of the diagram (Figure 3.27) [13].[2] The evolution of smart meters has been incremental, beginning with the integration of short-range RF technologies that allowed "drive-by" capture of meter readings,

[2] Best Networking and Communication Product Award.

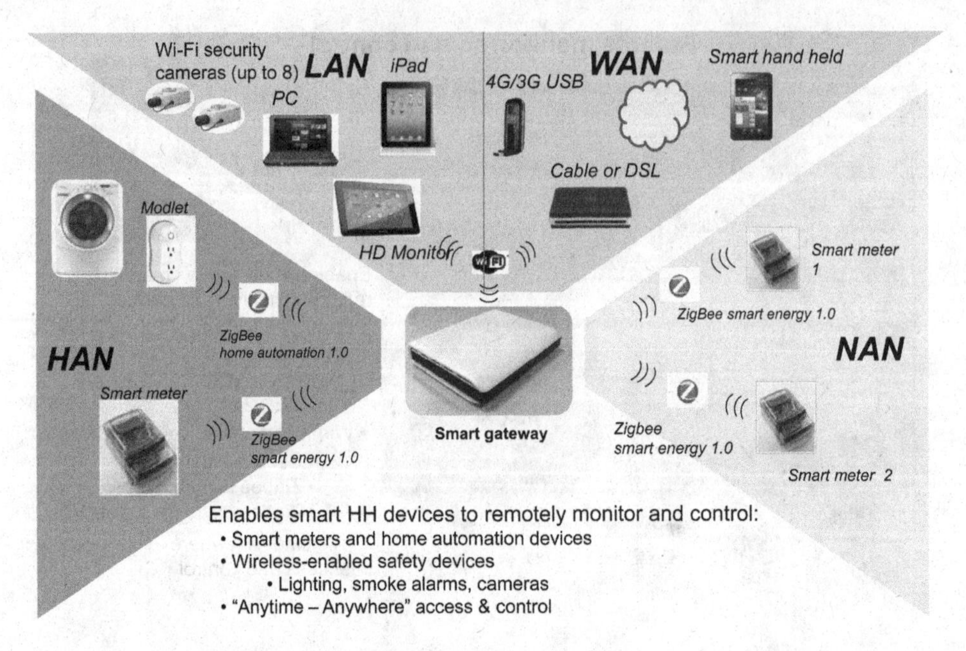

Figure 3.27 Networked smart gateways multiple services

which saved time and improved accuracy. Next, smart meter improvements followed that eliminated the need for a mobile field staff to capture meter data. This latest round of smart meter improvements has been based on standards-based communication technologies driven by the AMI organization. The use of standards-based communication technologies has opened the door for existing home networking platforms, such as residential gateways and broadband AP routers, to incorporate connectivity support. This new class of platform is referred to as a "smart energy gateway," or alternatively, a "networked smart gateway (NSG)" and represents the heart of the HAN.

3.1.5.3 Key Benefits of Smart Energy Gateway or NSG

A smart energy gateway provides the necessary interface between the utility-controlled smart grid and energy consuming in-house object. The following are some key benefits:

- Control activation/deactivation of HAN appliances
- Collect real-time energy consumption from smart meter and power consumption data from various in-house objects
- Generate dashboards to provide feedback about power usage
- Provide control menus to control appliances
- Connect to cloud; WAN for remote monitoring and control.

3.1.5.4 Challenges in Implementing Smart Energy Gateway

The key modifications necessary to enable a residential gateway to serve as a smart energy gateway is support for the PHYs and protocols that have been adopted for use in smart meters and home automation devices. The PHYs used in smart meters include: HomePlug Green PHY (Power Line Communication), ZigBee technology, and 802.11. The associated communication protocols include DLMS, Smart Energy 1.0, Smart Energy 2.0, and M-Bus in the European market. The integration of these PHYs together with support for communication protocols associated with them is a key requirement for designing a smart energy gateway. In parallel, these same PHY technologies are rapidly being adopted for use within the HAN to support home automation connectivity with appliances, lighting, security systems, and health monitoring devices, as illustrated in Figure 3.27. AMI as well as the AHAM, the USNAP consortium, and HEMS Alliance, are guiding this enablement. Other considerations that must be factored into smart energy gateway designs include backhaul support to WAN, and perhaps most importantly, a UI that enables access, monitoring, and control over the connected HAN devices.

This last factor deserves special emphasis. The ability to access, monitor, and control devices within the HAN is an essential capability if consumers are going to successfully manage and conserve the energy they use. The UI must allow customers to see exactly how much energy they are using, how much the utility company is charging for that energy, and provide the ability to exercise control over HAN-connected devices, if necessary.

Of equal importance to providing this UI is the ability to access it remotely, at any time, via a smart handheld device or tablet. This capability leverages the global preference by consumers to utilize a single smart device for all their communication applications, extending from voice, texting, email, and entertainment to now include home monitoring, security, and control.

3.1.5.5 Smart Energy Gateway

The smart energy gateway solution (Figure 3.25) is designed to address the challenges mentioned above. This multifunctional energy gateway can support M2M connectivity from smart handheld devices, such as smartphones, or tablets, to HAN. Through this M2M connection, a user can remotely monitor energy usage, receive alerts from their utility company regarding billing or tariff changes, and manage the smart devices within his/her HAN. Due to this, it serves as a single-chip solution providing all necessary secure connections between end-to-end devices in smart grid network.

3.1.5.6 Smart Energy Gateway Functionalities

The smart energy gateway (Figure 3.27) with integrated ZigBee module enables connectivity with any ZigBee-enabled HAN device. This allows the HAN to be accessed and controlled over an M2M link between a smart handheld device and NSG, using an intuitive GUI. The smart energy gateway supports HAN, WLAN, and WAN connectivity, with an integrated MPC13226 ZigBee radio for HAN connectivity to smart meters as well as smart plugs and appliances, an integrated 802.11n Wi-Fi module that delivers 300 Mbps of WLAN performance and support for broadband WAN connectivity via either cable, DSL, or LTE/3G. In addition, NSG also supports the latest 802.11n Wi-Fi radio modules via its miniPCI connector, as well as dual Gigabit Ethernet ports, to enable true broadband connectivity.

For wireless Internet connectivity, a 3G or 4G USB module can be attached via either of the two high-speed USB 2.0 ports that are provided on the NSG.

The smart energy gateway reference design kit includes a complete suite of OpenWRT software that requires no license fees and supports the following applications: GUI that enables Web-based access and management of connected devices and applications, NVR for home surveillance, HD video streaming, and ZigBee HAN profiles for Smart Energy 1.0 and Home Automation 1.0.

3.1.5.7 Key Features of Smart Gateway

- *Seamless wireless connectivity (TCP/IP, 802.11n, ZigBee).* Smart metering via ZigBee sensors (via SE 1.0 or MBus)
  - Remote management and control of smart appliances (via ZigBee HA1.0)
  - M2M "anytime/anywhere" access and management, via smart handheld or Web-enabled devices
- *Simple Web-GUI.* Easy to use with any Web-enabled device; meter reading, energy consumption, and history alert notifications of tariff changes by utilities in real time
  - DR: Manage energy usage (HVAC, lighting, car charging, etc.)
  - Home automation and security
- *Integration of four essential software stacks.*
  - TCP/IP: Broadband WAN/LAN connectivity
  - ZigBee Home Automation 1.0 Profile
  - ZigBee Smart Energy 1.0 Profile
  - Dual-PAN radio that supports both HA1.0 and SE2.0 simultaneously on one ZigBee radio
  - Web-based GUI (Java)
- Cost-optimized bill of materials
- Enables "anytime/anywhere" access and control over an M2M link from any smart handheld device via its GUI
- Provides a superb price/performance blend and the horsepower to run a variety of applications simultaneously
- Supports a rich mix of networking capabilities such as VoIP; HD video streaming, home security, and surveillance; and energy management, and home automation control

- CE and FCC Class A certified, RoHS compliant, ready for mass production
- Complete OpenWRT software suite.

3.1.5.8 Summary on HAN (Home Area Network) Smart Gateway

Bridging the smart grid with the HAN, the networked smart gateway solution delivers new possibilities in home energy monitoring, while allowing utility companies to tailor specific energy packages. The multi-application versatility and cost-effectiveness of the networked smart gateway makes it an ideal solution for adding home energy management and control capabilities to a standard broadband gateway platform.

Smart Home Gateway allows one to remote control appliances using mobile smartphones and tablets. This will empower consumers, help them change their behavior, and reduce their bill.

In addition, a large increase in electricity demand is expected in the coming years, as traditional energy electricity is replacing other sources; this is especially true for cars, where electric vehicles are considered a more environmental friendly solution for the future.

To be able to accommodate all these anticipated changes, the electricity grid needs to transition from a hierarchical, unidirectional, and centralized grid to a distributed and networked grid accepting injections of power generated by consumers using renewable energy resources.

To ensure power availability and power efficiency, and to avoid power grid instabilities resulting from bidirectional power flows and less predictable demand, integrated information and communication technologies (ICT) network acting as a control plane is also required.

The bidirectional, decentralized, electricity grid, and its associated ICT control plane are at the heart of the smart grid, but will however only be able to maximize its benefits if communication to the home and among appliances within the home can be ensured.

The promises of the smart grid can therefore come to fruition only if greater response, greater engagement, and active participation from end consumers within homes can be ensured.

Active consumer participation in DR program will enable users to contribute to grid flexibility and resilience, and enable users to adjust their electricity demand (and thus their bill) as a response to price signals or reliability-based actions.

Greater engagement in distributed energy resources will offer consumers an unprecedented array of choices in how to use, store, or sell their energy in line with their economic and social values.

The above-mentioned smart energy services and applications, enabled by the smart electricity grid will however not be hosted on the smart meter, the grid's endpoint into residential homes, but will need to be hosted on complementary devices like the home energy management box. This modular device will help provide comprehensive management of energy within customer premises.

3.1.5.9 Home Energy Management Gateway (HEM)

Most consumers only have a vague idea of the amount of energy they use for different purposes [14]. They are however very much concerned about their energy bill and are willing to save money, behave "greener," and save energy, provided they understand where the electricity goes, how much they waste, and provided they understand how they can derive tangible benefits from an optimized consumption.

It is therefore important for consumers to be able to track energy costs in real time, and to be able to understand the various energy flows within the home or building.

The smart meter makes it possible for consumers to save on their energy bill, but does not do that by itself; smart meters mainly help utility companies get better readings on electricity use and help utility companies save energy and money.

To help consumers do the same, while at the same time making sure that the utilities' expectations regarding consumer adoption rate in DR or load management programs are met, it appears that besides smart meters, rollout of consumer-friendly companion devices within homes is mandatory.

The home energy management gateway connected to a multitude of devices within the home, ranging from smart meter, smart washing machine, remote controllable HVAC, to PHEV charging station, is one such consumer-friendly device. It can host automatic DR programs reacting to utility messages according to user-set policies, but also host nonintrusive load disaggregation software making the end users aware of their real time energy consumption pattern and behavior.

The 24/7 always-on gateway can help solve the problem of premises energy management and either act as an energy coach or take a more active role and act as a dynamic load shifting controller. The energy coach can propose energy saving tips aimed at reducing the overall energy profile, and the load shift controller can silently manage one's home or small business energy usage, and sequence nonsimultaneous activation of controllable devices to low energy cost times, on an hour-by-hour or even minute-by-minute basis.

In the future, the gateway could also act as an energy manager coordinating energy flows within premises: according to real-time time-of-use electricity rates, local generation capabilities and local electricity storage levels, the box could balance the user's comfort level and lifestyle with a minimized utility bill and optimize local energy use/store/sell decisions.

A typical deployment scenario for the Home Energy Management Gateway within the home is shown in Figure 3.27.

The Home Energy Management Gateway receives price events or demand–response events through the AMI network and its smart meter interface to the home (or potentially through the broadband interface).

- Monitors and controls a set of demand–response-enabled appliances (e.g., thermostat, water boiler, and heater)

- Monitors and controls a set of home-automation-enabled appliances (e.g., washing machine and dimming lights)
- Reads power figures out of the various meters and loads on a periodic basis
- Serves rich analytics (e.g., load disaggregation results) to various online, mobile, or local displays, and is ready to
- Runs certified third-party value-add widgets (e.g., energy saving widgets)
- Monitors microgeneration unit production
- Monitors electric vehicle charging
- Sequences controllable loads, so they do not run simultaneously or sequences soft-start gradual recovery mechanisms after power outage helps optimize both electrical power generation and transmission losses as lower current needs will also result in lower quadratic-dependent wire losses.

3.1.5.10 Home Energy Management Development Platform

The Home Energy Management reference platform has been developed [8] as an open hardware and software development platform aimed at jumpstarting design of new products complementing the smart meter rollout.

The platform can serve as the basis for devices targeting consumers willing to manage and control, in real time, consumption of electricity and other energy loads in a building or a house.

The platform comes with prevalidated technologies enabling household or building occupants to remote control or program an array of automated devices, to manage energy usage in an intelligent way, and to interconnect with the different technologies in the home.

Targeted first for energy management, time of use, or DR-ready devices, the platform has been envisaged such that upcoming requirements for deployment of value-added applications (such as customer-specific tips and hints for optimizing energy consumption, utility incentives), detection/prediction of appliance and equipment failures, video surveillance, monitoring of microgeneration units or distributed storage, or electric vehicle charging, can easily be addressed in the future.

The platform is powerful enough not only to collect real-time data delivered by a network of intelligent meters and sensors but also to run a framework processing and delivering analytical visualizations in an intuitive and powerful visual way through a number of access methods such as touch screens, mobile phones, and web browsers.

As such, the platform provides a solid foundation for the development of feature-rich and easy to install HAN appliances supporting exciting ways of engaging, educating, motivating, and empowering individuals to conserve resources and save money.

Optimized to address today's market needs, the Home Energy Management development platform is modular and designed with the future in mind by accommodating for upcoming and foreseeable requirements such as monitoring microgeneration units or electrical vehicle charging. Extension ports and connectors, for example, power-line modems, RS-485-based solar inverters, and additional communication modules are also available on the platform.

Additional value-added software, in the form of an OSGi® framework or a home automation framework, is available from ecosystem partners and enables customers to further reduce their overall design cycle and their overall time to market.

3.1.6 Industrial and Factory Automation

Much of our transport, civic, and industrial infrastructure (e.g., road, rail, tunnels, bridges, waterways, and pipelines) asset base is still monitored and maintained using very labor-intensive processes. M2M technology has real potential to provide more cost-effective autonomous techniques to help with remote monitoring and preventive maintenance. While their needs often differ, they face one common problem: the cost involved in cleanup operations and insurance payouts after subsidence, landslip, or breached canal embankments is usually huge. While open spaces lend themselves well to using wireless technologies, it is not uncommon to find fiber laid along the lengths of rail-track, roadways, and so on, that could form part of the solution infrastructure (see Figure 3.28). Key selection criteria for the appropriate embedded processors include connectivity, storage, and packet processing, in order to support M2M gateways in infrastructure. Figure 3.29 shows the industrial IoT connectivity model for Smart industrial, building, and factory automation as well as for retail and Point of Sale (POS) and ATM kiosks. Table 3.7 shows the software stacks for the industrial IoT.

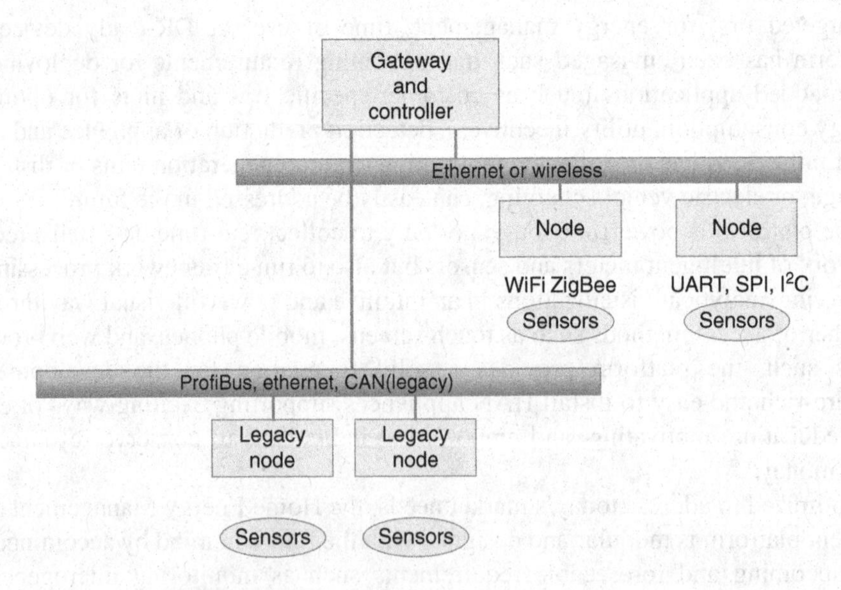

Figure 3.28 M2M industrial infrastructure

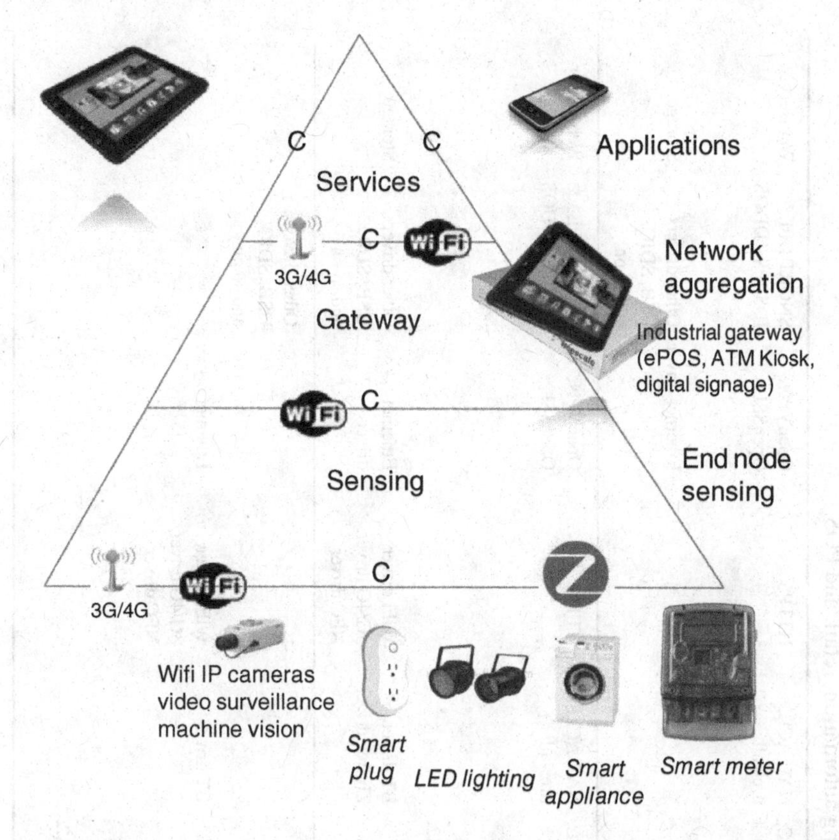

Figure 3.29 Smart industrial, factory/building automation, retail, and POS/ATM

3.1.6.1 M2M for Smart Manufacturing, Smart Factories

Smart manufacturing includes factory automation such as automated assembly and packaging using robotics, machine vision, and machine learning. Smart industrial gateways are used to connect legacy machines with numerous legacy connectivity platforms such as CAN bus, Ethercat, and Profibus. Programmable Logic Controllers are used in SCADA network of this industrial equipment [15].

3.1.6.2 M2M for Smart Retail

Smart retails include deployment of ePoS, ATMs, kiosks, smart vending machines at shops and hospitalities. Deployment of digital signage and self-service kiosks and checkouts is growing rapidly. The addition of sensors and short-range wireless to digital signs has turned them from screens into interactive displays, providing stimulus for advertisers who can now target and connect with new customers. NFC technology allows users to interact with advertisements via their smartphones.

Table 3.7 Software stacks for smart industrial, factory/building automation, retail, and POS

	CoAP	DB – SQL, unstructured DB	Security, IPSEC, SSL, DTLS, encryption	NAT, ACL, firewall	HTTPS	OpenVPN, OpenSSL	SNORT DPI, IDS/IPS – DDOS	Web-server
Service (cloud)	Networking, ipv6, VLAN, DNS	Cloud apps – billing, and so on				Ethernet driver	Other drivers, USB, SDIO, and so on	Storage
	CoAP	DB – SQL, unstructured DB	Security, IPSEC, SSL, DTLS, encryption, secured boot	NAT, ACL, firewall	HTTPS	OpenVPN, OpenSSL	SNORT DPI, IDS/IPS – DDOS	Web-server
IoT gateway, (wireless router)	Networking, ipv6, VLAN, DNS	Provisioning and Mgt, TR69, QOS	Wireless security, WPA2, WEP					
				BT driver, ZigBee driver	WiFi driver, 3G/4G driver, NFC driver	Ethernet driver	Other drivers, USB, SDIO, and so on	Storage
							Other drivers, USB, SDIO, and so on	
	ZigBee driver		Security, IPSEC, SSL, DTLS, encryption, secured boot	BT driver	WiFi driver, 3G/4G driver, NFC driver	Ethernet driver		
Sensing node			Wireless security, WPA2, WEP					

Context-aware and location-based services (LBS) IoT leverages sensor technology can be used to identify context and demographics (who, where, when and in the future, the mood) of passersby and deliver context-aware services accordingly. Owing to its connectivity and high-performance features, general-purpose applications can be added, such as digital signage capabilities.

3.1.6.3 Summary on Smart Industrial (Grid/Energy, Buildings, Factories, Retail)

While the particulars of each use case may vary, the rationale for using M2M is consistent: safety, security, power, and cost savings.

The potent combination of advanced packet processing, energy management, and integrated I/O flexibility with trusted boot will allow developers and integrators to quickly deploy wired and wireless M2M systems they can trust. When considering the volume of M2M nodes and gateways which will be deployed, often in reasonably accessible places, trust and security are vital.

3.1.7 Smart Transportation and Fleet Logistics (Connected Cars – V2X: V2V, V2I)

Smart transportation applies to smart car, smart bus, smart train, and so on. The automobile industry already has implemented some M2M in the past. An example is the TPM (tire-pressure-monitor), which has a pressure sensor integrated with an MCU that detects tire pressure at each tire and transmits the reading to the instrument panel also known as DIS (Driver Information System). The DIS also has numerous operational sensor information of the car such as speed (odometer), RPM reading, motor, and transmission oil level, water, and coolant fluid level, GPS location, brake pad wearing, which are connected to the OBD II (On board Diagnostic) connector underneath the steering column as well as the display on the instrument panel. After-market products can be purchased that take the information from the OBD II connector and send it to a smartphone in the car, which can then send the same to the cloud (Internet). One use-case of this is that insurance companies can allow drivers to select an option where their driving habits can be monitored remotely and the insurance companies can provide a discount to drivers with good driving habits.

The car's OnStar device detects a crash (air bag deployed) and automatically sends the GPS location and owner's information to an emergency response center which dispatches 911 assistance.

Many cars already are equipped with ADAS (Advanced Driver Assistance System) such as active braking and active speed control that detect your car is approaching too close to another object, then decelerate the car by apply braking, lane change warning that detects your car has drifted away from your lane, blindspot detection detects objects in your blindspot, and so on.

Smart transportation is tracking of vehicles such as truck and taxis with GPS tracking connected to the cloud so that real-time fleet logistics and scheduling can be performed.

Additional OBD II data such as speed can also be transmitted to the cloud in real time so that the trucking and taxi companies can monitor their drivers to ensure safe driving practices.

Figure 3.30 shows the IoT connectivity model for Smart Transportation and Table 3.8 shows the software stack for Smart Transportation.

An emerging inter-car communication application being developed is the smart connected cars with V2X (Figures 3.31 and 3.32) for active safety and accident avoidance. Information on vehicle-to-vehicle (V2V) communication and vehicle-to-infrastructure communication (V2I) can be accessed from http://www.Dot.gov. In the next few years, each new car equipped with V2V will have a wireless smart gateway that transmits an 802.11p WiFi DSRC (Digital Short Range) beacon consisting of GPS location, speed, and direction of travel while receiving beacons from other cars in the vicinity. The smart gateway in your car will compute in real time with low latency to warn you that you will have a collision in 10 s if you continue on the same path at the same speed. Your onboard smart gateway can also trigger active speed control and active braking to take action as well as relating collision and traffic jam messages to the cars behind you once your car receives a collision notification from the car in front. With inter-car communication, accident messages can be deployed

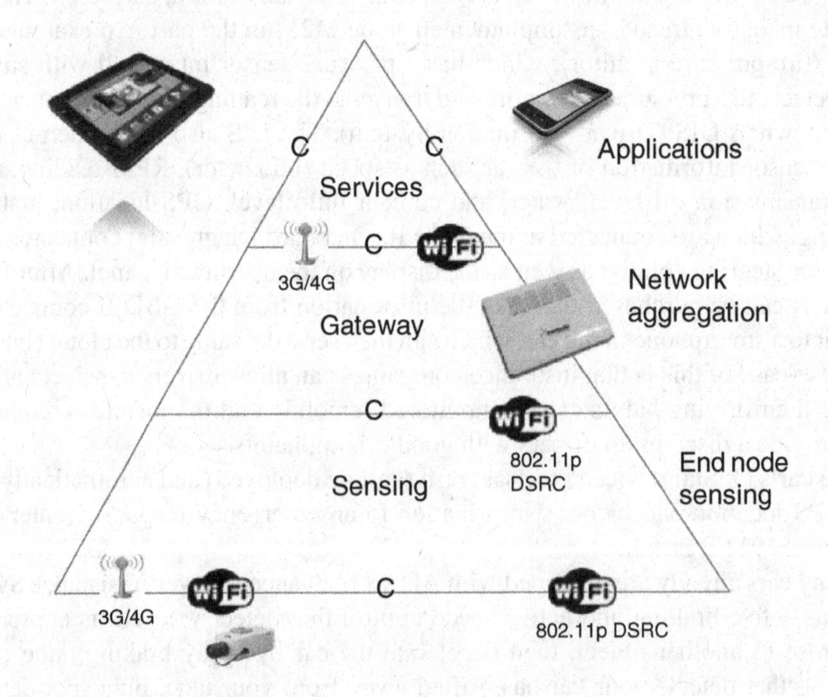

Figure 3.30 Connected smart transportation, smart car – V2X (V2V, V2I)

Table 3.8 Software stacks for connected smart transportation

	CoAP	DB – SQL, unstructured DB	Security, IPSEC, SSL, DTLS, encryption	NAT, ACL, firewall	HTTPS	OpenVPN, OpenSSL	SNORT DPI, IDS/IPS – DDOS	Web-server
Service (cloud)	Networking, ipv6, VLAN, DNS	Cloud apps – billing, and so on						
						Ethernet driver	Other drivers, USB, SDIO, and so on	Storage
	CoAP	DB – SQL, unstructured DB	Security, IPSEC, SSL, DTLS, encryption, secured boot	NAT, ACL, firewall	HTTPS	OpenVPN, OpenSSL	SNORT DPI, IDS/IPS – DDOS	Web-server
IoT gateway, (wireless router)	Networking, ipv6, VLAN, DNS	Provisioning and Mgt, TR69, QOS	Wireless security, WPA2, WEP					
				BT driver, ZigBee driver	WiFi driver, 3G/4G driver, NFC driver	Ethernet driver	Other drivers, USB, SDIO, and so on	Storage
							Other drivers, USB, SDIO, and so on	
	ZigBee driver		Security, IPSEC, SSL, DTLS, encryption, secured boot	BT driver	WiFi driver, 3G/4G driver, NFC driver	Ethernet driver		
Sensing node			Wireless security, WPA2, WEP					

Smart connected cars –V2X (V2V, V2I)

- Active safety, accident avoidance (V2V, V2I)

> Wifi peer-to-peer (P2P) mesh networking

802.11p for V2V (vehicle-to-vehicle), V2I (vehicle-to-infrastructure)

– Car tracking (location, speed, etc.); logistics: taxi, truck, etc.

– Autonomous car (google car)

Figure 3.31 Smart connected cars

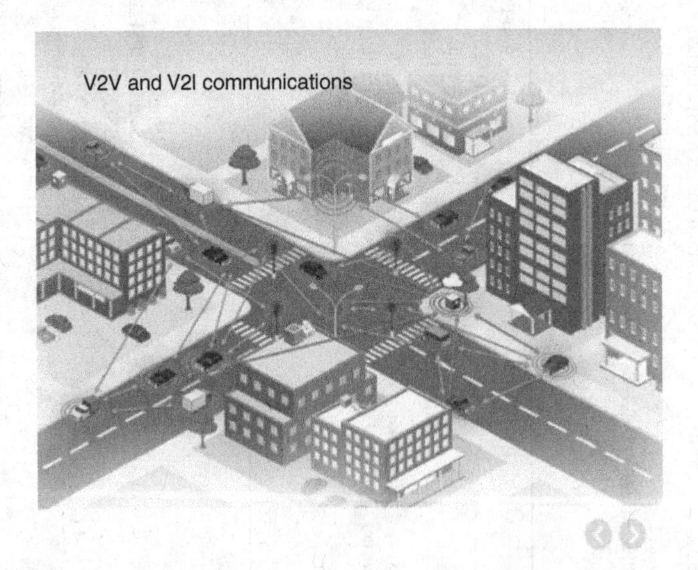

Figure 3.32 V2X (V2V and V2I)

in real time to cars behind, and your car can also take action to reroute in order to avoid the traffic jam and collision ahead.

V2I application allows your on-board smart gateway sensors to communicate with the road infrastructure such as traffic lights and speed signs. If your car is following a truck or in a fog, you will be alerted if you attempt to drive through a red light when it is changing from green to red. Likewise, you will be alerted if you are driving over the speed limit posted by the speed limit sign and conceptually, once this road infrastructure is cloud connected, a ticket can also be issued to you and charged to your RFID tag automatically as well.

In V2X, data received can also be traffic information such as congestion, accident, service reminders, targeted marketing, and sales information. Data transmitted by V2X can be vehicle health data regarding engine, transmission, emission (such as OBD II), driver performance data, anti-theft information, telematics data such as GPS location.

3.1.7.1 Smart Transportation (Car, Bus, Train, etc.) Vehicle Telematics

OBD II information is commonly used by vehicle telematics devices that perform fleet tracking, monitor fuel efficiency, and prevent unsafe driving, as well as for remote diagnostics and by pay-as-you-drive insurance. Although originally not intended for the above purposes, commonly supported OBD II data such as Vehicle Speed, RPM, and Fuel Level allow GPS-based fleet tracking devices to monitor vehicle idling times, speeding, and over-revving. By monitoring OBD II DTCs, a company can know immediately if one of its vehicles has an engine problem and by interpreting the code and the nature of the problem. OBD II is also monitored to block mobile phones when driving and to record trip data for insurance purposes.

Ultimately, semi-autonomous and autonomous car with and without the driver will soon arrive.

3.1.7.2 Smart Traffic – Smart Roads, Highways, Road Infrastructure

Sensors can also be installed on roads and highways to provide guidance to the drivers such as speeding, crossing stop signs, and red lights. Sensors can be used to track traffic flow so that dynamic traffic light control can be implemented to optimize traffic flow to minimize traffic jams, resulting in fuel saving and reducing carbon emission that causes air pollution.

3.1.8 Smart City

Smart City infrastructure deploys street lighting control and video surveillance for public security, safety as well as parking meters, smart road/traffic, emergency response, and so on (Figure 3.33).

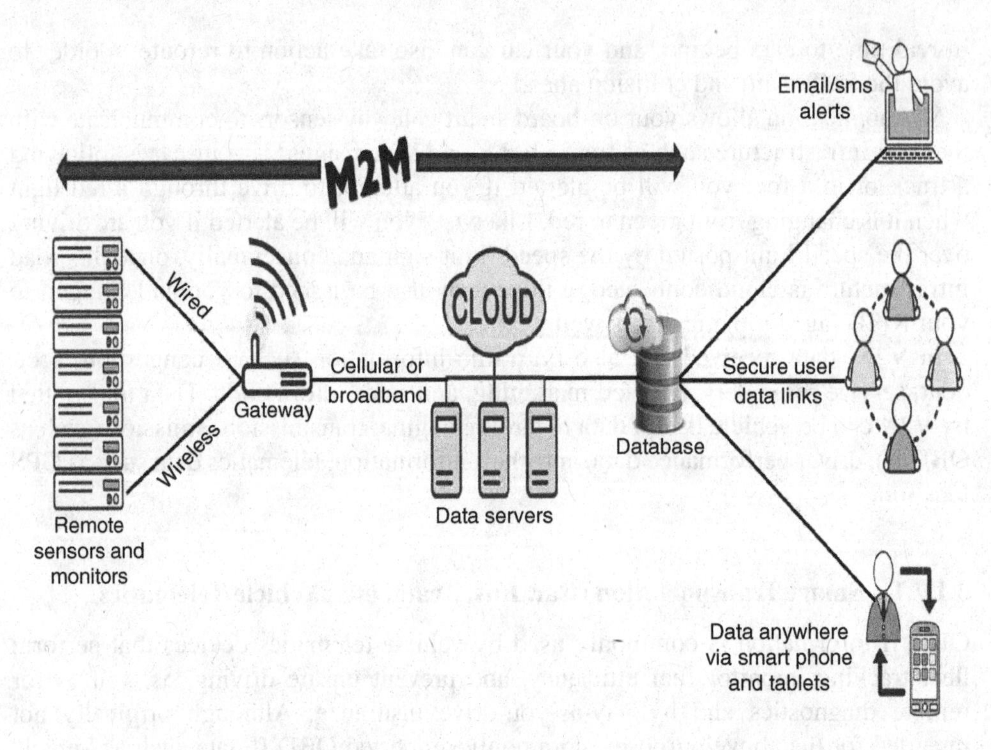

Figure 3.33 Smart environment in smart city

In addition, Smart City maintains smart environment by deploying sensors to monitor air and water quality, energy, waste, noise and road/traffic monitoring such as from Envirologger http://www.envirologger.com/.

In addition to building automation such as LED lighting, AC temperature, humidity control, sensors are also deployed for sensing structural issues of buildings and bridges, so that preventive care can be deployed before a major collapse happens.

In addition to smart street lights, video surveillance cameras are added for public safety and security, as well as smart traffic monitors of vehicles and pedestrian traffic and optimizes the flow of vehicles and pedestrian routes. Smart parking meters notify drivers on available parking spots. Smart transportation systems include smart roads and highways with warning messages for weather, accidents, and traffic jams.

3.2 Smart IoT Platform

This section presents the use of Smart IoT Platforms to deploy IoT products, systems, and solutions. There are 7S's success criteria for IoT Systems: the top 4S's of IoT success criteria are Simplicity, Secured, Smart software, and Scalable.

- *Simplicity*. Easy to use, install and reset, Easy on the Eyes, Simple Solution to Complex Problems
- *Security*. Privacy, protection from intrusion
- *Safety*. Safety, stability, supportable
- *Smart*. Adaptive, Analyze, Action, Anticipate (Location/Context Aware, Predict), Autonomous
- *Scalable service (HW and SW) and Standards (open-standards)*. Easily scalable by various types and number of end nodes and users (interoperate with other devices); as well as data rate
- *Sustainable, Performance, and Power efficient*. Long battery life, power saving modes, energy harvesting
- *Sleek Appeal and Aesthetic*. Good Look and Feel (Attractive) to match Smart Living, Smart Environment, and Community.

The following sections expand on two of these important success criteria – Smart and Secure IoT software platform.

3.2.1 Smart IoT Software Gateway Platform

This section expands on the "Smart" IoT systems that will prove an important success factor for mass IoT adoption [3, 5].

https://community.freescale.com/community/the-embedded-beat/blog/2014/07/16/smart-IoT-systems--the-missing-s

Smart IoT system solutions are based on "Smart" software platforms (Figures 3.34 and 3.35, Table 3.9) characterized by 7A's. The top 4A's are awareness of context, analyze and take action, anticipate, automate, and autonomous.

- *Awareness of Context and Location.* Smart IoT system solutions are customizable by context and location. Smart software behaves differently according to who, when, where, and so on. The rule-based abstraction provides additional simplification and ease of use.
- *Analyze and Take Action.* Smart IoT systems have balanced local storage that allows analysis, local processing, and data/event filtering at the sensor node and gateway level, as well as making localized decisions and taking actions at the cloud level. This enables faster response and lower latency rather than always going to the global cloud for action, which in turn results in more intelligent, selective transmission of sensor data so as not flood the cloud with Big Data.
- *Anticipate, Predict.* Smart software understands the user and knows their usage history. It then anticipates, making relevant predictions based on context and use history. This can help deliver targeted sales and marketing solutions and services.

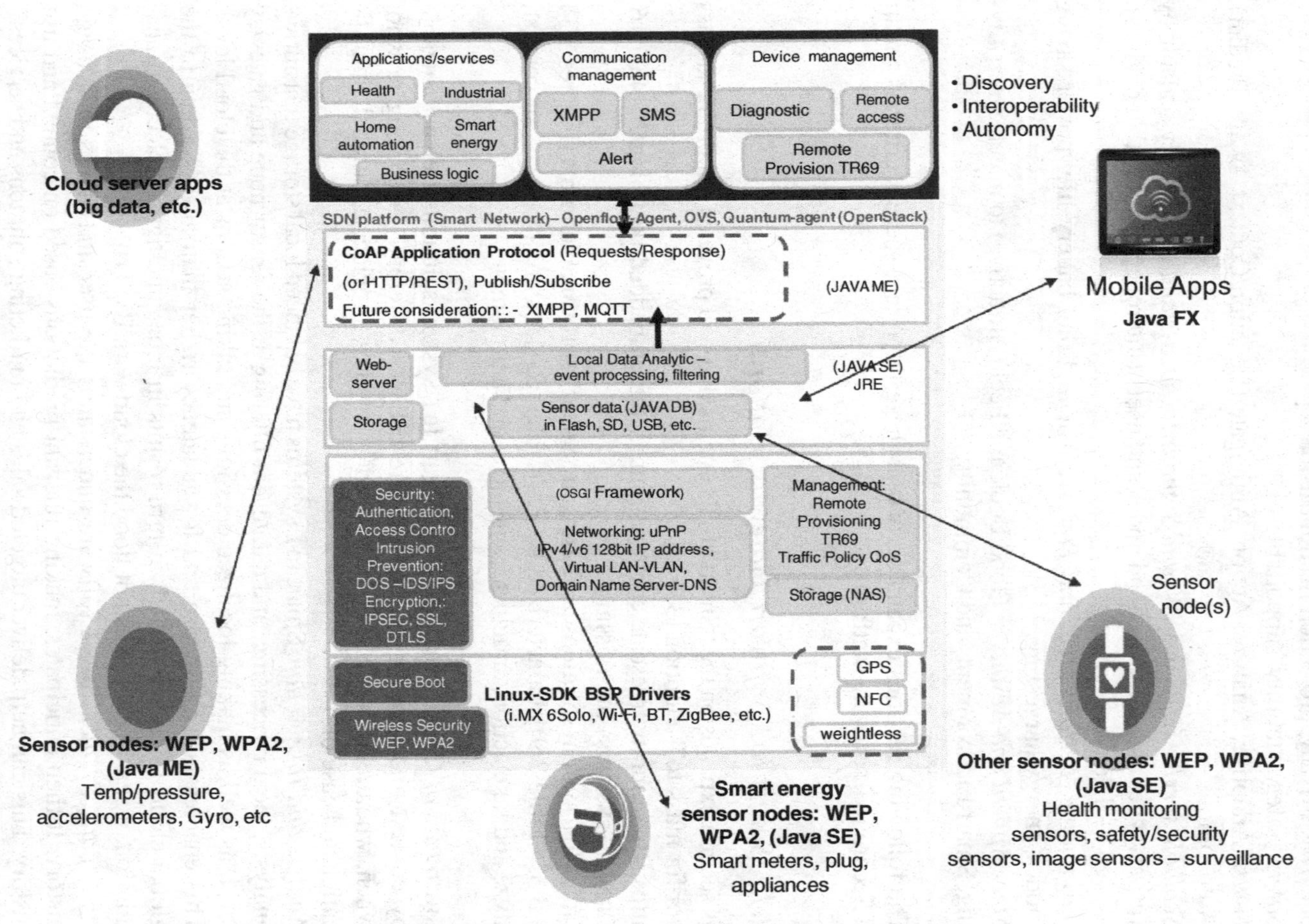

Figure 3.34 Smart, secured IoT software platform

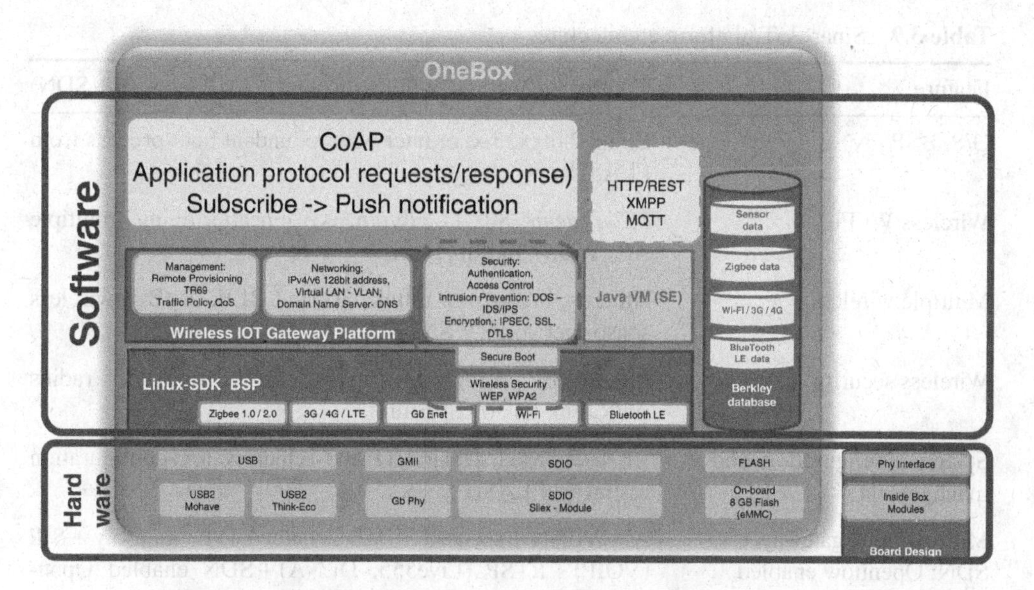

Figure 3.35 IoT software platform architecture

- *Automate.* Apply simple solutions to complex problems. Smart IoT products will improve efficiency by automating and streamlining processes through automated monitoring and control.
- *Autonomous.* A truly smart system independently makes informed decisions and takes appropriate actions with a self-governing, self-organizing ability. Autonomous systems are usually rule based with a knowledge-based reasoning ability. Smart devices can "discover" each other and can interoperate (collaborate) together. In addition, future smart devices and smart gateways will be connected through the Software-Defined Network (SDN). This facilitates the entire end-to-end network to be visible and manageable as an adaptive smart system, so high availability such as auto-failover and load balancing can happen dynamically to shift traffic load from one equipment to another.
- *Automated Remote Provisioning and Management.* Typically, this is cloud based, from initial installation and configuration to management. The ability to support remote monitoring, tracking, management, and control is critical. Smart sensor devices and smart gateways can and must be remotely managed, serviced, and sustained.
- *Augmented Reality.* This can take many forms, including easy-to-use HMI, gesture user interface, voice in/voice out with natural language analysis/interpretation, and data/knowledge mining. Adaptivity is also important. Products with learning mode and dynamic, on-demand real-time rule-based adjustment.
- *Aesthetic.* Sensory attraction is important too. Smart products must have a pleasant look and feel, while also delivering a strong user experience.

Table 3.9 Smart IoT platform architecture

Feature	Comment (Secured Wireless Router + IoT Gateway + SDN)
O/S, BSP	Linux 2.6.xx, 3.x, or later with redundant boot process from FLASH
Wireless Wi-Fi	*802.11acnn*, 802.11 a/b/g/n, AP repeat/bridging (multiple APs WDS), multiple SSIDs
Multiple wireless	Atheros, broadcom, ralink; 3G and BLE, ZigBee wireless sensor network (WSN)
Wireless security	PSK, WEP, WPA2, ipv4/ipv6 authentication with radius (w/Ralink RT3092 Wi-Fi)
System configuration and management	Browser based, TFTP: FTP and Telnet, system configuration restore in FLASH
Multi-service gateway + SDN: Openflow enabled	AP-WLAN + secured SMB router + IoT gateway + SIP (VOIP) + RTSP (Live555, DLNA) + SDN enabled Openflow agent + OpenVSwitch
L2/L3 routing Protocols + DPI (SNORT)	IPv4 and IPv6, RIP v1/v2, TCP/IP, UDP, PPTP, PPPOE, DMZ, VLAN, Bridging, NAT/PAT, IGMP, VPN, Firewall, ACL, IDS (SNORT DPI), VRRP
UPnP	Open source SDK with configuration interface and integration
DHCP	DHCP client, DHCP relay agent, DHCP server
DNS	DNS relay, dynamic DNS

Smart IoT software platforms include rich M2M connectivity (3G, WiFi, ZigBee, WSN), local storage, and cloud connectivity; they are equipped with rule-based automation that contributes to simplicity and user-friendly interfaces. Smart software abstracts hardware complexity, so the hardware is more transparent to the user and becomes simpler and easier to use.

The smart IoT gateway platform includes message-based RESTful API for synchronizing with cloud servers through "request versus publish" types of message-based interfaces (e.g., CoAP – Constrained Application Protocol or MQTT).

As mentioned by the author in a previous blog post – the "I" in IoT also stands for "intelligent" networks. Smart IoT gateways are also SDN-enabled with an OpenFlow agent that allows smart sensors and gateways to be visible and managed by a centralized OpenFlow controller.

Smart IoT systems based on the above smart IoT software platform will help accelerate the development of vertical market applications and rate of IoT adoption.

Figure 3.34 shows an IoT platform with Connectivity, Sensing, Gateway, and Cloud Services.

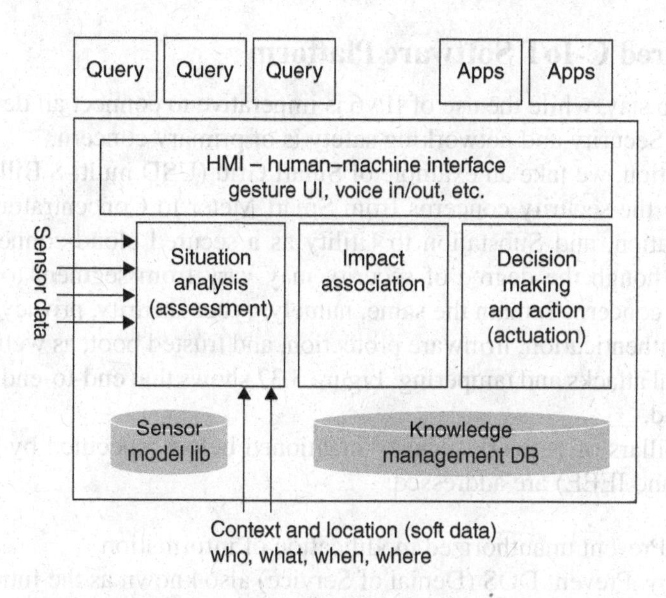

Figure 3.36 Sensor fusion software platform architecture

3.2.2 Smart Sensor Fusion Software Platform

This section introduces the concept of Sensor Fusion Platform (Figure 3.36). While more sensors are being integrated into microcontrollers to form smart sensor nodes, multiple sensors are still delivering sensing data that need to be processed and analyzed to result in decision-making and to trigger actions.

Sensor fusion takes hard sensor data fused with context-aware data such as location and owner data – referred to as soft data (who, what, when, where, etc.) – to perform situation assessment analysis; rule-based/knowledge-based decision making action can then be taken intelligently at the edge nodes. The C-IoT service platform leverages sensor fusion software framework components, to perform ubiquitous sensing and processing works transparent to the user. This will enable the development of C-IoT applications that interoperate among multiple point-solutions of Smart things. This will further drive the IoT market products to new heights.

The Sensor Fusion Platform architecture implements Smart senor analysis and decision-making [4]. Though traditionally implanted in the cloud with Big Data analytics, it is best to have distributed intelligence where some portions of the capabilities are distributed to local gateways and sensor nodes to minimize latency and improve response time.

The sensor-fusion platform consists of software modules that perform sensor data analysis, impact assessment, and knowledge-based approach decision-making. The framework is extensible, so new sensor data can easily be added to the sensor library and a new knowledge base can easily be added through learning. Additional data analytic applications can also be added to leverage the underlying engines and database.

3.3 Secured C-IoT Software Platform

IoT is here to stay, while the use of IPv6 is imperative to connect all devices globally via Internet. Security and networking safety is of primary concern.

In this section, we take an example of Smart Grid (USD multi-$ Billions markets) and describe the security concerns from Smart Meter to Concentrators, Concentrator to Substation, and Substation to Utility as a secured cloud-connected SCADA network. Although the degree of security may vary from segment to segment, the IoT security concerns remain the same, namely, cyber security, privacy, confidentiality, access authentication, firmware protection, and trusted boot, as well as protection from physical attacks and tampering. Figure 3.37 shows that end-to-end security must be considered.

All four pillars of network security mentioned below (specified by NIST SP 800 82, NERC, and IEEE) are addressed:

- *Integrity*. Prevent unauthorized modification of information
- *Availability*. Prevent DOS (Denial of Service) also known as the Intrusion Prevention System.
- *Confidentiality*. Prevent unauthorized access of information
- *Non-repudiation*. Prevent denial of action.

For each of the device in the Smart Grid, we suggest the platform solution (hardware and software) that can be used to make it secure and reliable.

This section includes the distribution grid automation with improved monitoring and visibility of utility's huge distributed assets, which is one of the key motivations to provide significant return on investment (ROI) to the utility companies. For example, AMI with automated smart data concentrators and transmission line monitoring sensors can improve business operations such as DR to prevent brownout and does not require significant behavioral change by customers. The utility benefits are reduced downtime, faster service restorations, and rapid identification of faults for preventive maintenance.

The M2M and IoT end-to-end security solutions provided in this chapter can be extended to other areas like Smart Energy, Smart Health, Smart Transportation, Smart Factory, Business, and Residence (Smart Home) and make connected intelligence a reality.

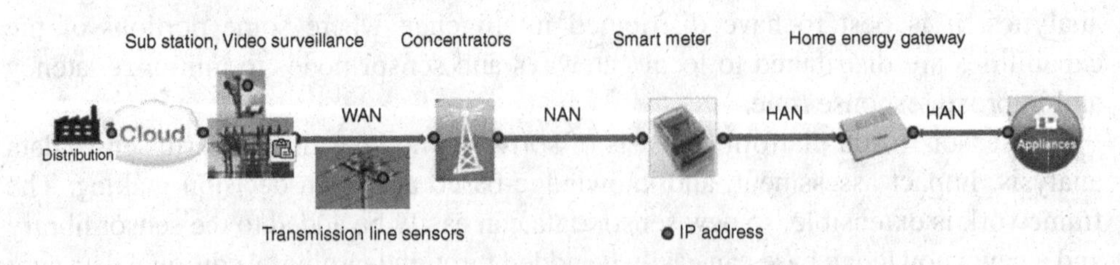

Figure 3.37 End-to-end security in the smart grid

3.3.1 Overview

In this section, we take a closer look at the security concerns at the communication layer, which predominantly, we believe, will be the IP-based network. As the data transmit over Internet, assured security in terms of Data Integrity, Confidentiality, and Non-Repudiation [16, 17] is essential. All devices must be protected against DOS Attacks, cyber attacks, vulnerabilities, and exploits. They also should be protected from malformed and bad traffic. Every IP address is susceptible to attacks and must be protected.

Smart Meters, Concentrators, and Transmission line sensors are deployed out in the open and thus must be protected from physical attacks and tampering. Such attempts have to be detected with the help of sensors and reported to the utility. Trusted boot and secure architecture must be provided to protect against unauthorized or malicious firmware upgrade.

Reliable two-way communication is essential for effective functioning of the smart grid. Having a mesh topology at the concentrator and at the substations can provide reliability and redundancy. Load balancing and failover (LBFO) can be achieved with two WAN connections at the concentrator and substations. For security, data are encrypted and for reliability RAID5 (redundant array of independent disks) can be implemented on upstream devices to protect loss of transient data.

We summarize the security concerns across the Smart Grid hierarchy as shown in Figure 3.38. Some examples of platform solutions, covering various aspects of security such as confidentiality, integrity, non-repudiation, and availability discussed are as follows:

- *Smart Meter.* Low footprint IP Stack with 6LoWPAN, Firewall, and IPSec-IKEv2 running on ZigBee SoC (System-on-chip) (Figure 3.39)
- *Concentrator.* IP Stack on Concentrator with 802.15.4 and 802.3 as in Figure 3.41
- *Home Energy Gateway.* With WiFi and Security Software (Figure 3.27)
- *Utility and Substation Servers.* With hardware accelerators for high performance.

3.3.2 C-IoT Security – Example of Smart Energy

Let us take a closer look at the security concerns at the communication layer of the Smart Grid which predominantly is expected to be the IP-based network as shown in Figure 3.37. As the data transmit over Internet, it has to be assured security in terms of Data Integrity, Confidentiality, and Non-Repudiation. All devices must be protected against DOS attacks, cyber attacks, vulnerabilities, and exploits. They also should be protected from malformed and bad traffic. Every IP address is susceptible to attacks and must be protected.

Smart Meters, Concentrators, Transmission line sensors are deployed out in the open and thus must be protected from physical attacks and tampering. Such attempts have to be detected with help of sensors and reported to the utility. Trusted boot and secure architecture must be provided to protect against unauthorized or malicious firmware upgrade.

Reliable two-way communication is essential for effective functioning of the Smart Grid. Having a mesh topology at the Concentrator can provide reliability and redundancy. Additional reliability of communication can be achieved by having two WAN connections at the Concentrator and Sub Stations for LBFO. Reliability can be further enhanced by implementing RAID5, Redundant Array of Independent Disks (https://en.wikipedia.org/wiki/RAID) on upstream device to protect loss of transient data.

Let us summarize the security concerns across the Smart Grid hierarchy shown in Figure 3.38 and propose platform solutions. Some examples of platform solutions covering various aspects of security discussed are:

1. *Smart Meter*. Low footprint IP Stack with 6LoWPAN, Firewall, and IPSec-IKEv2 running on ZigBee SoC.
2. *Concentrator*. IP Stack on Concentrator with 802.15.4 and 802.3.
3. Home Energy Gateway with WiFi and Security.
4. Utility and Substation Servers.

3.3.2.1 Securing Communication on the Smart Grid

This section covers the various layers of Smart Grid Hierarchical Architecture, security concerns for each layer, and platform solutions for each. Smart Grid security can be divided into the following segments/application blocks as shown in Figure 3.38:

1. Neighbor Area Network (Smart Meter to Concentrator)
2. Home Area Network (Appliances to Smart Meter or Home Energy Gateway)
3. Wide Area Network (Concentrator to Substation/Utility).

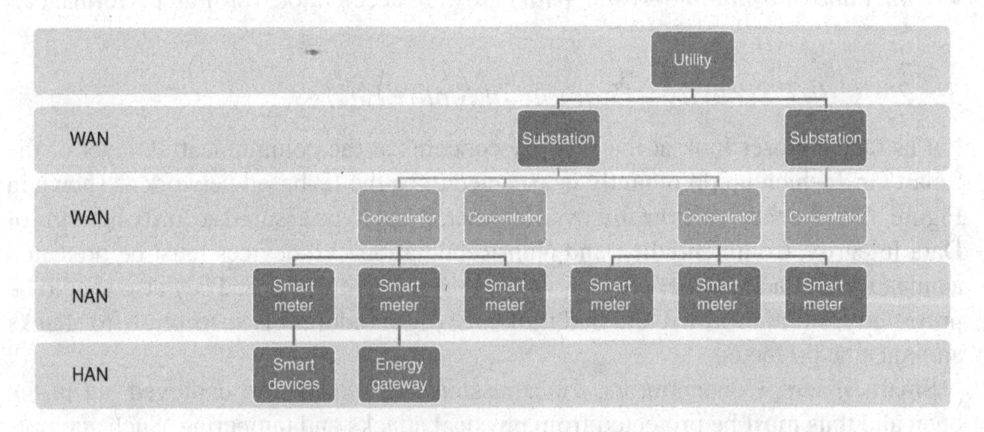

Figure 3.38 Smart grid hierarchical architecture

Each of these layers is described in the upcoming subsections in terms of their typical deployment, network topology, communication technology, security concerns for the devices in each segment, and protecting the data that are being communicated.

3.3.3 Securing NAN (Metrology-to-Concentrator)

NAN that connects the Smart Meter to the Concentrator is perhaps the most challenging in terms of vulnerability and security solutions for the following reasons:

1. Varied communication technologies options available are WiMAX, 3/4G, 802.15.4, and PLC. Although multiple technologies are available, most of them are wireless and suffer the drawbacks of wireless communication, where the data can be sniffed out and tampered with.
2. Low memory and other hardware resource constraints on the Smart Meter limit the feasibility of a robust security solution and restrict the options available for connection technologies.
3. Both Smart Meter and Concentrators are deployed outdoors and are susceptible to physical attacks.

These factors make NAN the weakest link in the Smart Grid and thus all factors impacting security need to be taken into consideration and addressed carefully.

This section takes 802.15.4 as an example of communication between Smart Meter and Concentrators and discusses the security issues and demonstrates how they can be addressed. The same can be applied to other modes of communication.

3.3.3.1 Platform Solution for Smart Meter

Since adoption of IPv6 is inevitable, in this example, 6LoWPAN/802.15.4 running on ZigBee SoC is proposed as a platform solution. The IP stack can be enhanced with rich security features integrating Firewall, IPSec [18], and IKEv2 [4] as shown in Figure 3.39. However, since the flash and RAM on the board are a low, lightweight version, the security application to fit into sub 100 K ROM and sub 100 K RAM is used. Further, VaultIC from InsideSecure (http://www.insidesecure.com/eng/Products/Secure-Solutions/Secure-solutions-products) can be used to secure the keys and certificates on the Smart Meter.

3.3.3.2 Network Topology and IP Addressing: NAN

For NAN, the Concentrator is considered to be the focal point, 802.15.4 coordinator, a Full Functional Device (FFD) and the Smart Meter, a Reduced Functional Device (RFD) with routing function. All Smart Meters in a given area or building are fully

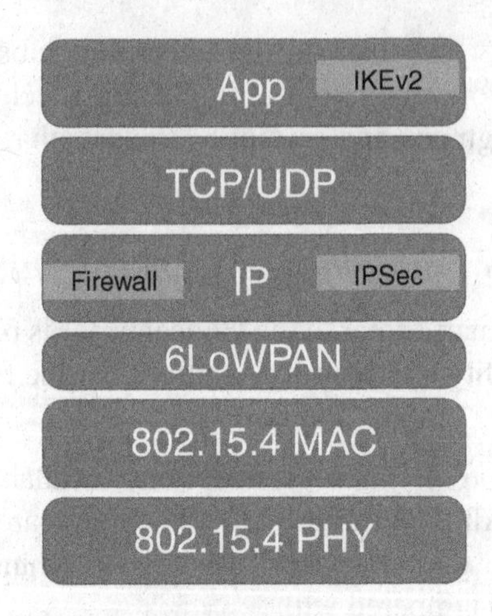

Figure 3.39 IP stack on smart meter with 6LoWPAN, firewall, and IPSec-IKEv2 on ZigBee SoC

meshed with the Concentrator as Coordinator. The Concentrator assigns DHCP v6 IP addresses to the Smart Meters.

3.3.3.3 Security on Smart Meter

The Smart Meter comes preloaded with shared keys or certificates [18, 19]. Alternatively, keys or certificates can be distributed in an out-of-band trusted mechanism. Since the IP address at the Smart Meter is dynamically assigned, an Internet Remote Access Client (IRAC) can be used for IKE exchanges. IKEv2 is used for Identity protection and for establishing Encryption and Authentication keys for the IPSec tunnel between the Smart Meter and Concentrator as shown in Figure 3.40. This combination of IKE and IPSec provides a secured NAN channel with data confidentiality and integrity. Sequence numbers in the IKE and IPSec help protect against replay and man in the middle attacks. Digital Signatures can be used for non-repudiation.

A list can be configured on the Firewall to allow only traffic from HAN to the Utility and vice-versa to pass through Smart Meter. Self-traffic can be limited only to allow IKE, DNS, DHCP traffic from the Concentrator to self and vice versa. Since the Smart Meter is expected to be actively communicating with the HAN devices, an access control list (ACL) should allow traffic from the HAN to Self and vice versa. DoS/Cyber-Attack check can be enabled on the Firewall to protect the Smart Meter from well-known attacks such as Ping Flood, Syn Flood, LAND Attacks, and Smurf Attacks. Any such attack detected should then be reported to Utility.

Secure Boot/Trusted Boot Architecture can be used to prevent malicious firmware upgrade. Anti-tampering sensors can be used to protect from physical attacks and tampering. Security concerns at the Smart Meter and the suggested solutions are summarized in Table 3.10.

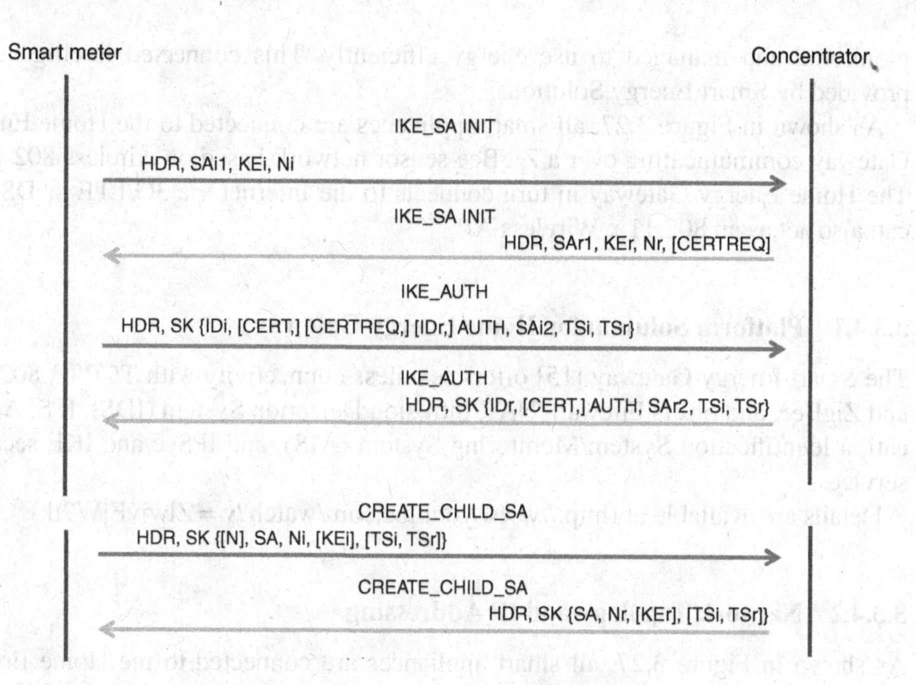

Figure 3.40 IKEv2 message exchange

Table 3.10 Summary of security on the smart meter

Security considerations	Proposed solution
Data integrity/authentication	IPSec/IKEv2
Data confidentiality/encryption	IPSec/IKEv2
Non-repudiation	IPSec/IKEv2
Replay/man-in-middle attacks	IPSec/IKEv2
Identity check	IPSec/IKEv2
Availability/denial of service	Firewall
Access control	Firewall
Trusted/secure boot	Secure boot arch
Anti-tampering	Anti-tampering sensors

3.3.4 Securing Home Area Network (HAN)

HAN has multiple smart automated appliances such as the HVAC, washing/dryer machines, smart plugs, lighting, and multimedia connected to Smart Meter and/or Home Energy Gateway; the Smart Meters from multiple households are then connected to the data concentrator in the NAN for AMI. These appliances can be monitored and managed to use energy efficiently. This connected intelligence is provided by Smart Energy Solutions.

As shown in Figure 3.27, all smart appliances are connected to the Home Energy Gateway communicating over a ZigBee sensor network based on wireless 802.15.4. The Home Energy Gateway in turn connects to the Internet via 3G/LTE or DSL. It can also act as an 802.11.x Wireless AP.

3.3.4.1 Platform Solution for Home Energy Gateway

The Smart Energy Gateway [15] offers seamless connectivity with TCP/IP, 802.11n and ZigBee, and offers Firewall, NAT, Intrusion Detection System (IDS), IPS, Application Identification System/Monitoring System (AIS), and IPSec and IKE security services.

Details are available at (http://www.youtube.com/watch?v = ZlwivEjW2tk).

3.3.4.2 Network Topology and IP Addressing

As shown in Figure 3.27, all smart appliances are connected to the Home Energy Gateway communicating over a ZigBee sensor network based on wireless 802.15.4. Smart metering connectivity is achieved via ZigBee SE 1 or MBus. Smart appliances are managed via ZigBee HA1.0.

On the WAN side, the Home Energy Gateway connects to Internet via 3G/LTE or DSL using DHCP, whereby the service provider dynamically assigns an IP address.

The wireless gateway can also act as an 802.11.x Wireless AP on the LAN side connecting laptops, tablets, and so on and has secured trusted boot architecture.

3.3.4.3 Security on Smart Energy Gateway

The remote monitoring, control, and management of all in-home Smart Appliances happens through the Smart Energy Gateway. These data are sensitive and private and thus have to be provided security while traversing the Internet. This is achieved by establishing an encrypted secure channel for this traffic over the WAN. IPSec/IKEv2 or SSL can be used to provide this. In addition to confidentiality, this solution provides integrity, identity protection, non-repudiation, and protection against replay and man in the middle attacks. Tight access control policies can be implemented using Firewall to allow only authorized traffic to and through the gateway. DoS/Cyber Attack check can be enabled on the Firewall to protect the Home Energy Gateway and the internal network from well-known D/DOS attacks. Deep Packet Inspection (DPI) software can be used to control/rate-limit application traffic, for example, P2P, Social Networking Application. Security concerns at the Home Energy Gateway and the suggested solutions are summarized in Table 3.11.

Table 3.11 Summary of security on the home energy gateway

Security considerations	Proposed solution
Data integrity/authentication	IPSec/IKEv2
Data confidentiality/encryption	IPSec/IKEv2
Non-repudiation	IPSec/IKEv2
Replay/man-in-middle attacks	IPSec/IKEv2
Identity check	IPSec/IKEv2
Availability/denial of service	Firewall
Access control	Firewall
NAT	Firewall
Application detection and control	Freescale AIS

3.3.5 Securing WAN (Concentrator-to-Substation/Utility Servers)

The communication between the Concentrator and the Substation can happen on one of several WAN technologies such as WiMAX, 3G/4G, or PLC. This communication is predominantly IP-based and data travel over the Internet. These sensitive bidirectional data that are from the Smart Meter to Utility and vice versa have to be protected from eavesdroppers to maintain confidentiality and integrity. As this link is critical to transfer real-time data, to ensure reliability it recommended having a failover connection. Critical messages from/to the meter must be prioritized over other traffic. The transient data on the concentrator have to be protected in case of crashes. Further, the Concentrator has to be protected from DOS attacks, bad traffic, and unauthorized access. As in the case of Smart Meters, since Concentrators are deployed outdoors, they have to be protected against physical attacks and tampering.

3.3.6 Platform Solution for Concentrator

The data concentrator [20] usually uses a low-power dual-core processor 667/800 MHz with up to 128 MB of NOR/NAND flash memory. The Security Accelerator, running Firewall, IPSec, IKEv2, and Application Identification Software can be used as a Concentrator as shown in Figure 3.41, which shows the data concentrator software stack delivering security and high performance.

This platform has 3G, WiFi, ZigBee WSN communication, and 3 Gb Ethernet capable ports to enable WAN/LAN communications and can communicate with Smart Metering devices via the industry standard DLMS (IEC 62056). It offers ZigBee wireless connectivity to meters and 3G Broadband to the Utility server. This device has an energy-efficient passive cooled design and has ruggedized, weather-resistant construction.

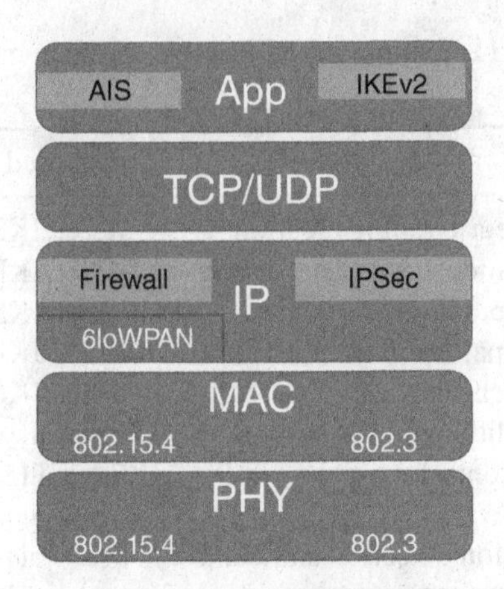

Figure 3.41 IP stack with 802.15.4 and 802.3 on data concentrator

3.3.7 Platform Solution for Substation/Utility Servers

The utility server is usually a multicore processor (>8 cores) [16] with Security Accelerator and Packet Matching Engine, running Firewall, IPSec, IKEv2, and AIS-Application Identification Software utility server delivers high performance with acceleration for the following functions:

1. Packet parsing, classification, and distribution
2. Queue management for scheduling, packet sequencing, and congestion management
3. Hardware buffer management for buffer allocation and deallocation
4. Encryption (SEC 4.0)
5. RegEx Pattern Matching (PME 2.0).

These devices can be "clustered" at the utility for load balancing and failover.

3.3.8 Network Topology and IP Addressing: WAN

The Concentrator's WAN interfaces get DHCP v6 addressed from the Substation. This connection can be 3G/LTE WiMAX, PLC. For greater reliability, two links to the Substation can be provisioned by LBFO software (Table 3.12).

3.3.9 Security on the Concentrator and Utility Servers

The Concentrator plays a key role in aggregating data from the Smart Meters within its area and propagating it upstream to the Substation or Utility over the WAN link. These data traveling over Internet have to be encrypted for confidentiality and privacy. A secure IPSec VPN-tunnel can be established between the Concentrator and the

Table 3.12 Security on the concentrator

Security considerations	Proposed solution
Data integrity/authentication	IPSec/IKEv2
Data confidentiality/encryption	IPSec/IKEv2
Non-repudiation	IPSec/IKEv2
Replay/man in the middle attacks	IPSec/IKEv2
Identity check	IPSec/IKEv2
Availability/denial of service	Firewall
Access control	Firewall
Vulnerability and exploit	Freescale AIS
Trusted/secure boot	Secure boot architecture
Anti-tampering	Anti-tampering sensors

Substation/Utility using IPSec/IKEv2. Since the IP address at Concentrator is dynamically assigned, IRAC can be used for IKE exchanges. This tunnel provides data confidentiality, integrity, non-repudiation, and protection against replay and man in the middle attacks.

ACL can be configured on the Firewall to limit traffic from HAN to the Utility and vice-versa to pass through the Concentrator. The ACL should allow active communication traffic from the Smart Meter to and from the Concentrator. Firewall can be enabled to prevent DoS/Cyber Attack on the Concentrator and report to the utility.

The DPI software can be used to protect the Concentrator fromr bad and malicious traffic, DPI has a rich set of signatures to detect bad traffic, P2P application, vulnerabilities, and exploits of protocols like HTTP, FTP, and SSH. On detection of bad traffic, action can be set to drop traffic and inform utility.

The Secure Boot/Trusted Boot Architecture can be used to prevent malicious firmware. Anti-tampering sensors can be used to protect from physical attacks and tampering.

Security configuration on the Substation and Utility server is similar to that on the Concentrator. In cases where the WAN IP address of the Substation is configured manually and known to the Utility server, site-to-site IPSec can be configured instead of IRAC. Although most equipment including the surveillance cameras are in premises and are relatively well covered from physical attacks, it is recommended to have anti-tamper sensors. Transmission line sensors and anti-tamper sensors should be used where the lines run above ground to detect and report attack.

Quality of Service (QOS) must be configured on the Concentrator, Substation, and Utility server to prioritize critical messaging data over other traffic.

3.3.10 Summary on C-IoT Security

The secure platform solutions proposed in this chapter facilitate distribution grid automation, which provides improved monitoring and visibility of a Utility

Table 3.13 Secure platform solutions for smart grid

Security concern	Home energy gateway	Smart meter	Concentrators	Substation and utility
	Proposed solution			
Data integrity/ authentication	IPSec/IKEv2	IPSec/IKEv2	IPSec/IKEv2	IPSec/IKEv2
Data confidentiality/ encryption	IPSec/IKEv2	IPSec/IKEv2	IPSec/IKEv2	IPSec/IKEv2
Non-repudiation	IPSec/IKEv2	IPSec/IKEv2	IPSec/IKEv2	IPSec/IKEv2
Replay attacks/man in the middle attacks	IPSec/IKEv2	IPSec/IKEv2	IPSec/IKEv2	IPSec/IKEv2
Identity check/data source authentication	IPSec/IKEv2	IPSec/IKEv2	IPSec/IKEv2	IPSec/IKEv2
Availability/denial of service	Firewall	Firewall	Firewall	Firewall
Vulnerability and exploit	Firewall+AIS	Firewall	Firewall	Firewall+AIS
Access control	Firewall	Firewall	Firewall	Firewall
Unauthorized firmware upgrade	Secure boot arch	Secure boot arch	Secure boot arch	Secure boot arch
QOS	VortiQa TM		VortiQa TM	VortiQa TM
Virtualization			KVM	KVM
Tampering, physical attacks	NA	Anti-tampering sensors	Anti-tampering sensors	NA

company's large distributed assets. Such remote monitoring capabilities serve as a key motivation for the Utility as it results in significantly improved ROI. For example, AMI with automated smart data concentrators and transmission line monitoring sensors can improve business operations such as DR to prevent brownout and does not require significant behavioral change by customers. Benefits to the Utility companies are: reduced downtime, faster service restorations, rapid identification of faults for preventive maintenance (Table 3.13).

3.3.10.1 Summary

This section has covered security concerns at various layers of Smart Grid and proposed end-to-end high performance security solutions viewing Smart Grid as a system. All four pillars of network security (specified by NIST SP 800 82, NERC, and IEEE) are addressed in the proposed platform solution:

1. *Integrity*. Prevent unauthorized modification of information
2. *Availability*. Prevent DOS and Intrusion.
3. *Confidentiality*. Prevent unauthorized access of information
4. *Non-repudiation*. Prevent denial of action.

The secured M2M, IoT solution provided here is not limited to the Smart Grid; this model can be adapted and extended to other applications/markets such as the following:

1. Gas and water distribution
2. Health, residence, and transport
3. Building and factory automation
4. Securing enterprise and data centers.

References

[1] Karimi, K. (0000) What the Internet of Things (IoT) Needs to Become a Reality, Freescale, http://www.freescale.com/files/32bit/doc/white_paper/INTOTHNGSWP.pdf (accessed 18 November 2014).

[2] Wu K. (0000) 2012 Top Embedded Innovator, Embedded Computing Design, http://embedded-computing.com/articles/2012-solutions-freescale-semiconductor/ (accessed 18 November 2014).

[3] Wu, K. (2013) http://www.imaps.org/chapters/centraltexas/Symposium_Presentations/Freescale_Internet_of_Things_IOT_Gateway_Platform_Next_Generation_KK.pdf, 2013.

[4] Karimi, K. (0000) The Role of Sensor Fusion and Remote Emotive Computing (REC) in the Internet of Things, Freescale Semiconductor, http://cache.freescale.com/files/32bit/doc/white_paper/SENFEIOTLFWP.pdf (accessed 18 November 2014).

[5] Wu, K. (0000) Blogs on Smart Internet of Things and Smart SDN Network, https://community.freescale.com/community/the-embedded-beat/blog/2014/07/16/smart-iot-systems – the-missing-s, https://community.freescale.com/community/the-embedded-beat/blog/2014/05/27/network-infrastructure-transformation-internet-of-things-iot-software-defined-networking-sdn-disruption-of-things-dot, https://community.freescale.com/community/the-embedded-beat/blog/2014/04/10/intelligent-integrated-solutions-to-bring-the-internet-of-things-to-life (accessed 18 November 2014).

[6] Niewolny, D. (0000) How the Internet of Things Is Revolutionizing Healthcare, Freescale Semiconductor, http://cache.freescale.com/files/corporate/doc/white_paper/IOTREVHEALCARWP.pdf (accessed 18 November 2014).

[7] Freescale (0000) Freescale Home Health Hub (HHH) Reference Design, Freescale, http://cache.freescale.com/files/32bit/doc/fact_sheet/HMHLTHHUBFS.pdf (accessed 18 November 2014).

[8] Winner of 2011 Broadband World Forum: Infovision Awardm – Freescale Secure Broadband Gateway (based on QorIQ) Speed Time to Market–Smart multi-Service Business Gateway, http://www.youtube.com/watch?v=5Ni0tbZk8aQ (accessed 18 November 2014).

[9] Freescale (0000) Freescale 8308nSG Smart Gateway, Freescale Semiconductor, http://cache.freescale.com/files/32bit/doc/fact_sheet/MPC8308NSEGFS.pdf (accessed 18 November 2014).

[10] Wu, K. (2012) Smart, Secure and Connected at Power.org, April 2012, http://www.power.org/events/connect-secure-and-speed-up-your-smart-networked-applications/ (accessed 18 November 2014).

[11] Wu, K. (2012) Smart Energy – NAN Data Concentrator. Embedded Design Magazine, 2012, http://embedded-computing.com/articles/2012-solutions-freescale-semiconductor/, www.youtube.com/watch?v=nI3Jb1LHL44 (accessed 18 November 2014).

[12] Wu, K. (2012) Internet of Things/M2M: Smart Converged Gateway for Smart Grid and Buildings, Oracle Java Embedded, October 2012, https://oracleus.activeevents.com/connect/sessionDetail.ww?SESSION_ID=12987 (accessed 18 November 2014).

[13] Wu, K (2012) Smart residential gateway (HAN), Australian and New Zealand Smart Metering Conference, Winner of 2012, http://www.youtube.com/watch?v=ZlwivEjW2tk (accessed 18 November 2014).

[14] Freescale Semiconductor (0000) Freescale Home Energy Gateway, Freescale Semiconductor, http://docbox.etsi.org/workshop/2011/201104_smartgrids/06_SGandTHEHOME/DUGERDIL_Freescale._HGI.pdf (accessed 18 November 2014).

[15] Davidson, I. (0000) Machine-to-Machine (M2M) Gateway: Trusted and Connected Intelligence, Freescale Semiconductor, http://cache.freescale.com/files/32bit/doc/brochure/PWRARBYNDBITSMTM.pdf (accessed 18 November 2014).

[16] Wu, K. and Rao T. (0000) Scalable Extensible Secured and Safe Smart Gateway Platform Solution for Smart Grid/ Energy and IoT, Freescale Semiconductor, http://link.springer.com/chapter/10.1007/978-3-319-03737-0_2 (accessed 18 November 2014).

[17] Balakrishnan, M. (0000) Freescale Security in Smart Energy, Freescale, http://cache.freescale.com/files/industrial/doc/white_paper/SESECURITYSEWP.pdf (accessed 18 November 2014).

[18] Freescale Semiconductor (0000) Freescale Secured Meter, Freescale Semiconductor, http://cache.freescale.com/files/microcontrollers/doc/fact_sheet/SECPPEMTRFS.pdf (accessed 18 November 2014).

[19] Balakrishnan, M. (0000) Freescale Smart Meter, Freescale Semiconductor, http://www.freescale.com/files/training_pdf/WBNR_SMARTMETER.pdf (accessed 18 November 2014).

[20] Freescale (0000) Freescale Security in Smart Meter, Freescale Semiconductor http://cache.freescale.com/files/industrial/doc/white_paper/SECSMTMTRWP.pdf (accessed 18 November\hb 2014).

[21] Freescale (0000) Freescale Smart Grid, Freescale Semiconductor http://cache.freescale.com/files/industrial/doc/brochure/BRSMRTENERGY.pdf (accessed 18 November 2014).

4

IoT Reference Design Kit

This chapter shows an IoT Reference Design Kit based on i.Mx6 processor that one could purchase online from CompuLab [1, 2].

http://utilite-computer.com/web/home

http://www.freescale.com/webapp/sps/site/prod_summary.jsp?code=RDIMX6SABREBRD&fsrch=1&sr=3&pageNum=1

Three models are available: single core Solo, Dual-core, or Quad-core.

The software image to download onto the board from www.freescale.com.

The IoT Reference Design Kit will provide the hands-on opportunity for one to experience the smart IoT software platform and IoT applications in delivery of various IoT products.

The Quick Start Guide provides the necessary step-by-step instructions to set up and demonstrate the IoT capabilities on the i.Mx6 board.

The software platform includes a Wireless access point (AP) Gateway that is easy to use and integrated with optimized open-source applications that support the use of wireless smart phones and smart tablets to perform remote monitoring, control of internet of things, and cloud services.

The following wireless-sensor-network (WSN)-based application can be demonstrated:

Home/Building Automation. Remote monitoring and control of appliances using a ZigBee-based Smart-plug modlet.

In addition, WiFi wireless-based media streaming applications can also be demonstrated:

Video Surveillance (NVR). Network Video Recording for video surveillance supports for wired or wireless IP cameras with auto detection.

Digital Living Network Alliance (DLNA) Server. Stream different video and music files from the wireless gateway board to multiple DLNA clients (e.g., Google TV, iPad, iPhone, and iPod touch).

Collaborative Internet of Things (C-IoT): For Future Smart Connected Life and Business, First Edition.
Fawzi Behmann and Kwok Wu.

4.1 Hardware Equipment List for the Demonstration

1. Hardware Platforms
   (a) Utilite i.Mx6 board or **MCIMX6Q-SDB Sabre board.**
   http://utilite-computer.com/web/home
   http://www.freescale.com/webapp/sps/site/prod_summary.jsp?code=RDIMX6SABREBRD&fsrch=1&sr=3&pageNum=1
   (b) SD Card containing boot image.
2. Wireless Smart Gateway Demo Peripherals
   (a) ThinkEco ZigBee USB dongle
   http://utilite-computer.com/web/home
   http://www.freescale.com/webapp/sps/site/prod_summary.jsp?code=RDIMX6SABREBRD&fsrch=1&sr=3&pageNum=1
   (b) ThinkEco ZigBee Modlet[1]
   http://www.amazon.com/ThinkEco-TE1010-Modlet-Starter-White/sim/B00AAT43OA/2.
   http://www.walmart.com/ip/ThinkEco-TE1010-starter-ThinkEco-Modlet-Starter-Kit-TE1010/21130985
   (c) Load for modlet (e.g., light and heater)
1. Media Streaming Demo Peripherals
   (a) WiFi IP Cameras – Axis Model M1011 (two sets)
   http://www.amazon.com/Axis-0301-004-0301004-M1011-W-camera/dp/B001PF3SDA.
   (b) Thin clients for DLNA, *optional, not provided* – iPhone or iPad with **Fusion Stream** app installed, iPod Touch with **iMedia Suite** installed, or Android tablet with **Skifta** installed.
2. Additional Peripherals
   (a) USB hub with external power (if using Mohave or 3G dongle in addition to ZigBee dongle)
   (b) 3G Broadband wireless Dongle – Huawei E160 O2 3G Dongle
   (c) AT&T SIM Card
   (d) Power Extension Cords (2)
   (e) Quick Start Guide and Poster
   (f) Large display monitor (user-provided)
   (g) Desktop or laptop computer with either the **Microsoft XP** (32-bit) or **Window 7** Professional (32-bit) operating system, user-provided.

4.2 Software Required for Demonstration

Software pre-flashed/pre-loaded on SD Card

Boot image
Video files in.mp4 or H.264 formats (for DLNA demo)
Smart-plug, smart-meter.

[1] Current modlets are rated only for 110 V, please use voltage adapter for 220/240 V source.

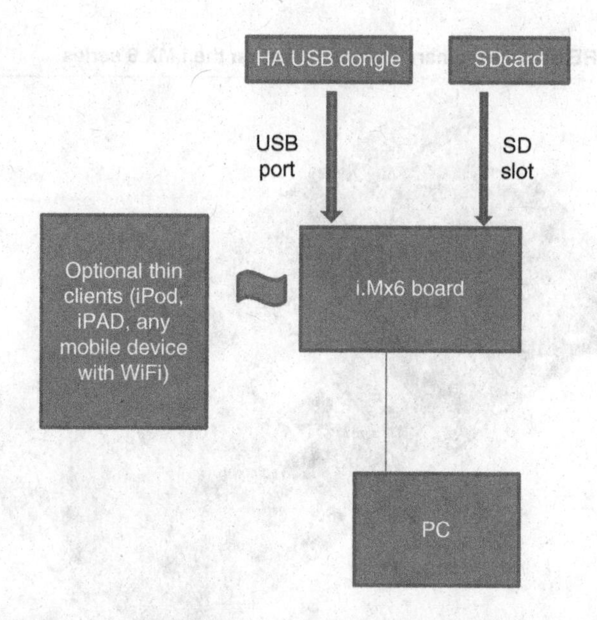

Figure 4.1 MCIMX6Q-SDB Sabre connectivity options

Installed Programs on PC

VLC for Windows (for NVR demo)
Axis IP Utility (for troubleshooting NVR)
Windows Media Player 12 (for DLNA demo)
TeraTerm (for troubleshooting).

To set up the I.MX6 Reference Platform (Figure 4.1):

1. Set up the Hardware according to Figure 4.2.
   (a) Power up the reference platform.
   (b) Configure the IP address of the network card of the Windows PC. Go to Start → Control Panel → Network and Internet → Network and Sharing Center.
   (c) If connecting to the reference platform wirelessly, select Wireless Network Connection. If connecting via Ethernet, make sure that the Ethernet cord is plugged into both the PC and reference platform's LAN port. Select Local Area Network (Figure 4.3).
   (d) Click on Properties → Internet Protocol Version 4. If connecting to the reference platform wirelessly, select "Obtain IP address automatically." If connecting via Ethernet, select "Use the following IP address," and enter static IP address 192.168.1.54 (Figure 4.4).
   (e) If connecting wirelessly, connect to the reference platform's SSID "I.MX60" using the password "12345678."

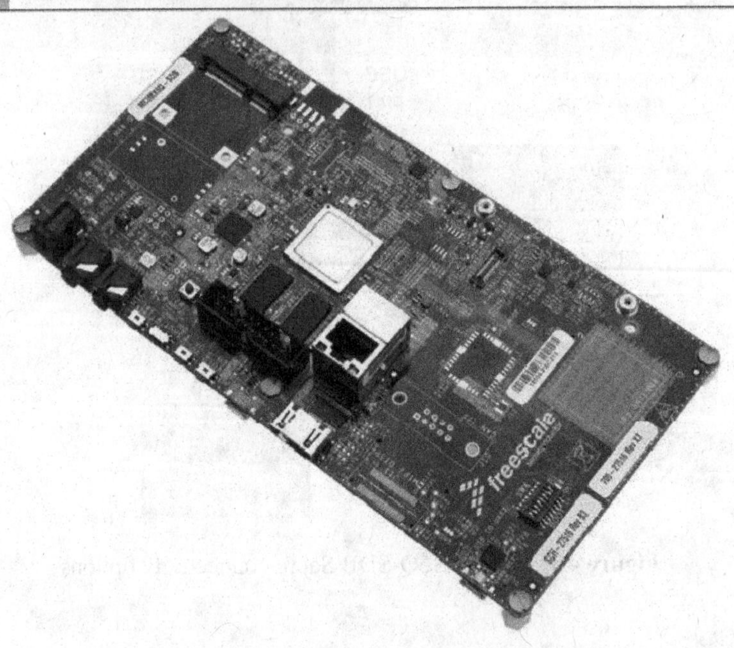

Figure 4.2 Basic setup diagram for I.MX6 board

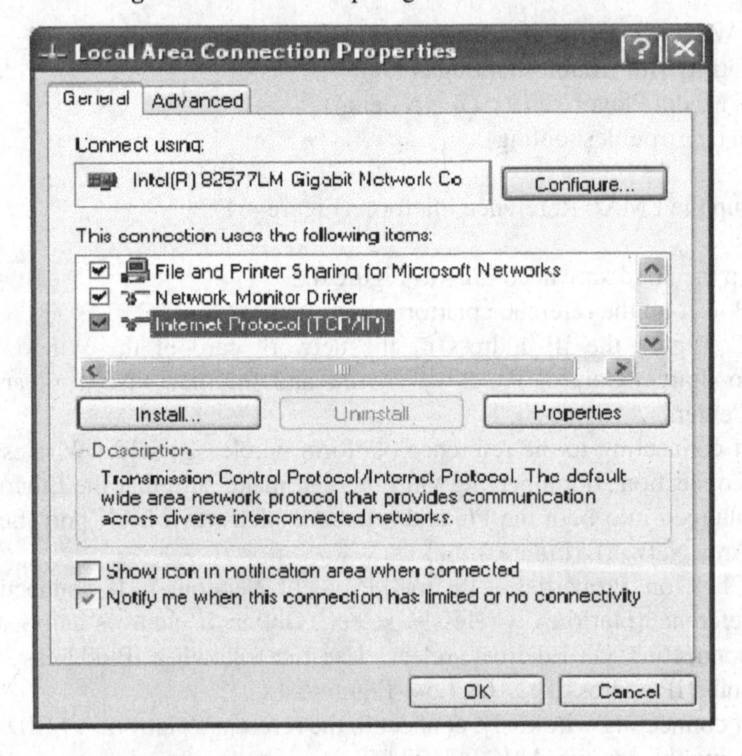

Figure 4.3 Local area connection properties

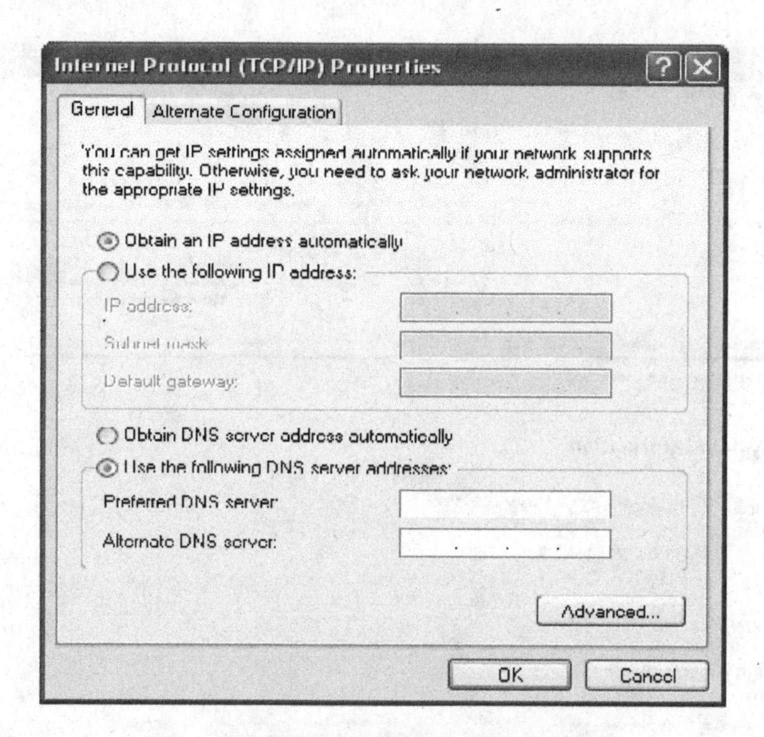

Figure 4.4 TCP/IP properties

(f) Open the Internet Explorer (IE) browser, ensure that you have no proxies set and enter 192.168.1.1.

(g) This opens the **reference platform** homepage; you will be prompted to set up/provide the username and password for future use.
  (i) Username: **root**
  (ii) Password: **root**

(h) Check Remember the username and password checkbox for future use and press Enter. The webpage is shown as shown in Figure 4.5:

1. If the SSID of the board is not "I.MX60" or if there is no password, set these up using the reference platform homepage as shown in Figure 4.6:
   (a) Select Network → Wireless.
   (b) For "SSID," enter "I.MX60."
   (c) For "Password (PSK)," enter "12345678."
2. If having trouble connecting wired devices to the box via an Ethernet switch, make sure that the Ethernet interface has been added to the bridge.
   (a) Use a computer to connect to the one box from "telnet 192.168.1.1" or TeraTerm.
   (b) Enter "ifconfig" to display interface configurations. Take note of the name for the Ethernet interface (e.g., eth0).

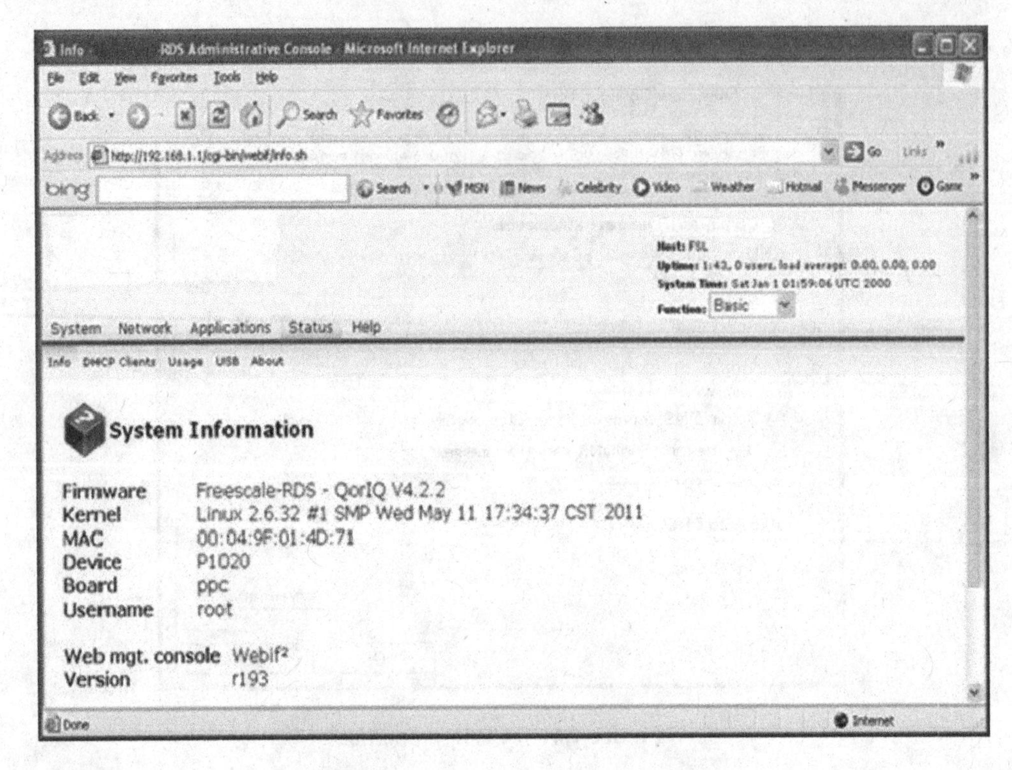

Figure 4.5 Reference platform homepage

(c) Enter "brctl show" to display bridge information.
(d) If the Ethernet interface is not displayed in the bridge information, add it using the command "brctl addif br-lan eth0" and the correct interface name found in Step 2.
(e) Enter "brctl show" again to make sure that the Ethernet interface has been added.

4.3 Safely Power Off the Reference Platform

To power off the reference platform safely without the loss of data or without corrupting the files in the mounted devices, follow the following steps:

1. Connect the serial cable to the SPI interface of the board and the COM port of the PC
2. Connect to the board using TeraTerm
3. Hit enter to go to the "**root@FSL#**" prompt
4. Type "poweroff."

Now, the board can be powered off.

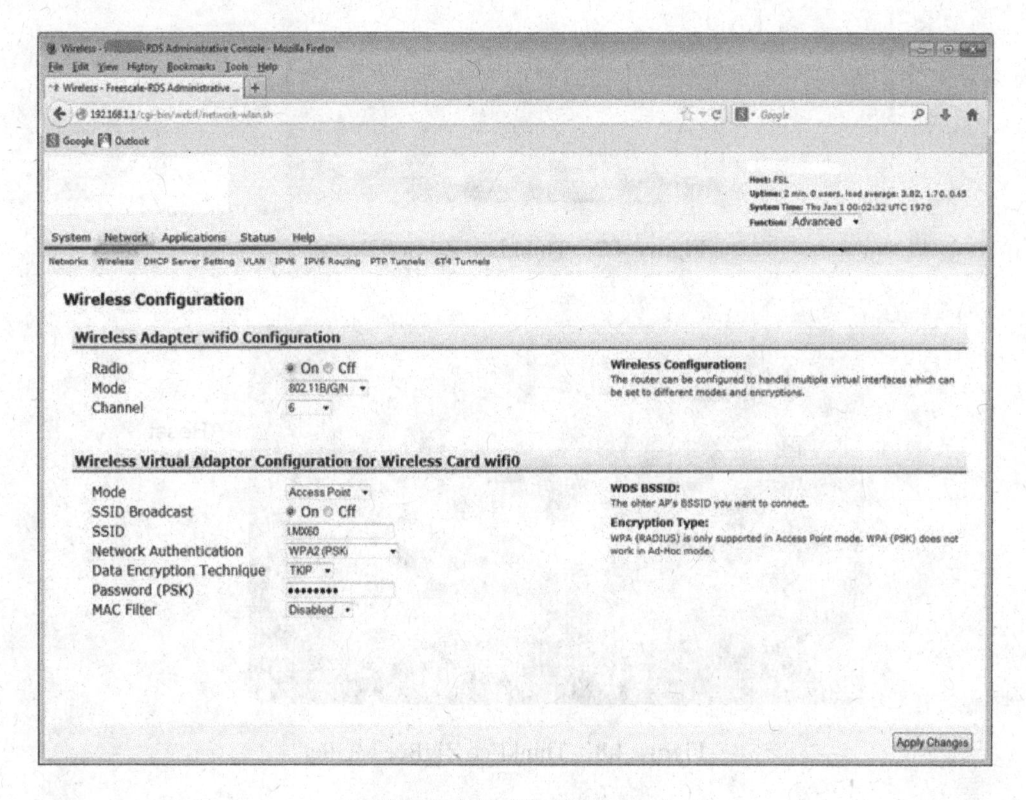

Figure 4.6 Wireless setup

4.4 ZigBee Home and Building Automation

This section describes the procedure to demonstrate the remote control of appliances through ZigBee WSN (e.g., Remote Power on/off device and monitoring power consumption of each appliance). Some applications examples are LED Lighting control, Smoke Alarm, garage-door opener, door-locks, window-shades for home, building automation, and so on.[2]

1. Insert the ThinkEco ZigBee dongle into the USB port of the reference platform as described in Figure 4.7.
2. Connect the **ZigBee Modlet** (smart plug) to a multi-socket power strip connected to the power meter and connect any electrical appliance (e.g., Light, Washer, and hair dryer) to the Modlet as shown in Figure 4.8.
   (a) Open Mozilla browser; type 192.168.1.1/home.html (Username: **root**, Password: **root**).
   (b) It will pop up message "click OK and press the reset button on the modlet."

[2] The home.html page can ONLY be open in ONE browser window at a time while running this demo. Only use Firefox, it does not work on IE.

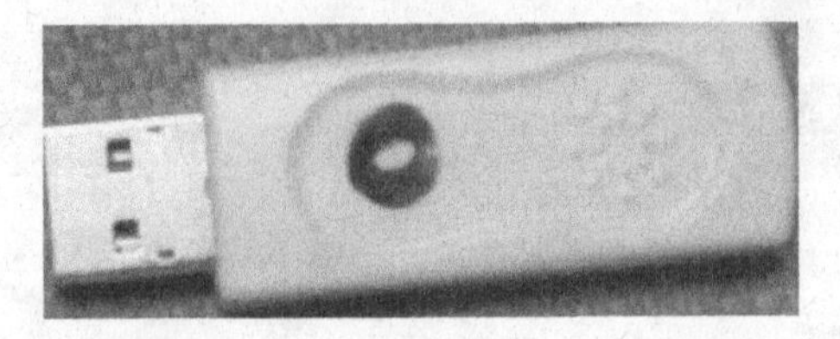

Figure 4.7 ThinkEco ZigBee Dongle

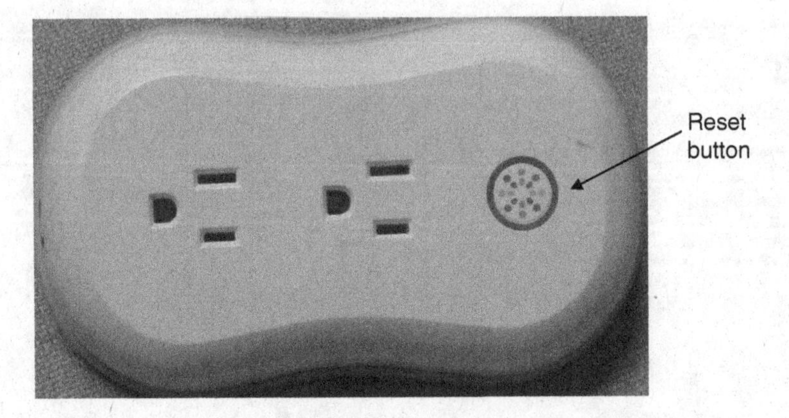

Figure 4.8 ThinkEco ZigBee Modlet

(c) Click OK while pressing the "reset" button on the Modlet (Smart plug) at the same time and hold it for 8–10 seconds until the red blinking light around the reset button turns off.
(d) If the light does not go off or if there is a message on the response page that said "Invalid message," refresh the browser page and repeat the previous step until you get the correct acknowledgment. Note: sometimes, the correct message will pop up even if it did not connect. The light on the modlet is the best indication. If it is dark, then it has connected. Anything otherwise means it did not connect properly.
(e) *Wait for 30 seconds after the permit-join before trying to toggle the power on or off.* To turn on the appliance connected to the upper socket (socket close to the button) of the Modlet, click "ON" in the backyard light pop-up window. To turn off the appliance connected to the upper socket (socket close to the button) of the Modlet, click "OFF" in the backyard light pop-up window.
(f) The power consumed by both appliances will be displayed in the same pop-up window as well as main screen.
(g) To perform the above-mentioned three tasks for the appliance connected to the lower socket of the Modlet, select "Dryer" instead of "backyard light."

4.4.1 *Troubleshooting ZigBee Home and Building Automation*

- If the smart plug does not respond to toggling the power on or off, make sure that home.html is only open in one browser and you are toggling the correct outlet. Refresh the home.html page and repeat the permit-join.
- If encounter "no dongle detected," the USB-fixed user space assignment for the dongle is incorrect. As a quick fix, remove the dongle from the USB hub and plug it directly into the board. As a permanent fix, get the updated 10-ftdi.rules file from the software download website and copy it to the SD card under the/etc/udev/rules directory.

4.5 Network Video Recorder (NVR) for Video Surveillance

This section describes the steps required to demo the NVR with auto detection of cameras.[3]

1. Ensure that the VLC player is installed, and Firewall and Proxy connections are disabled.
2. Open the IE browser and type **192.168.1.1** on the address bar.
3. Go to Applications → Network Video Recorder → Configuration page.
4. If the status of "Running" is No, click on **Restart NVR** and wait for restarting to finish (Figure 4.9).
5. Connect the real IP cameras to the i.Mx6 board.
6. Go to Applications → Network Video Recorder → Camera.
7. Click **Auto Detect Camera**, Cameras in the network will be automatically attached to NVR system. If no cameras are detected, follow the instructions in Appendix C to flash the camera (Figure 4.10).
8. Check the status of the cameras that are automatically detected by **Applications → Network Video Recorder → Status**. Make sure that all the visual cameras and real cameras are attached.
9. To watch camera videos, go to **Applications → Network Video Recorder → Channel** on the homepage.
10. Select **3 × 3** and select the camera to watch.
11. Click on the camera that you do not want to watch. Remove the check mark and click on
12. Save Changes.
13. In the pop-up window, click **OK**. 3 × 3 video stream windows should be shown.
14. If you want to make one video full screen, just double click on the video. If in one video full screen state, double click the video will return to the 3 × 3 video stream.

[3] Please install VLC for Windows. *This must be done in IE ONLY.*

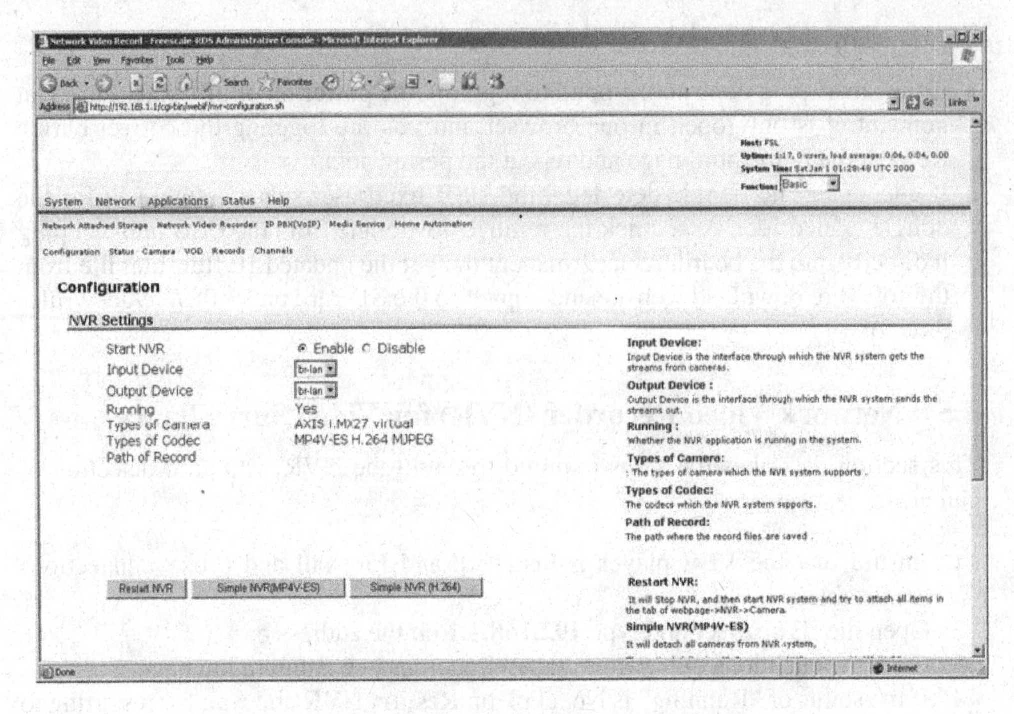

Figure 4.9 NVR configuration

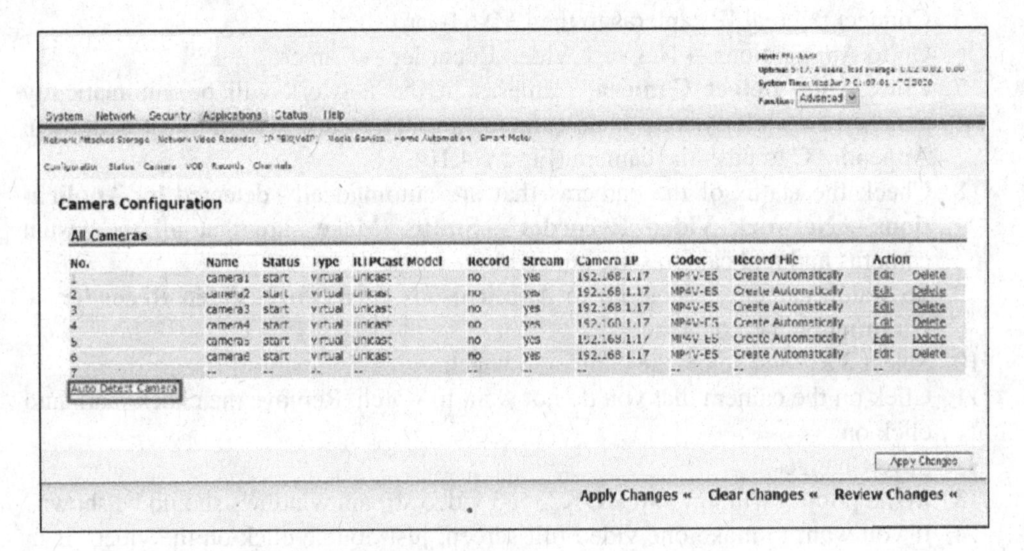

No.	Name	Status	Type	RTPCast Model	Record	Stream	Camera IP	Codec	Record File	Action	
1	camera1	start	virtual	unicast	no	yes	192.168.1.17	MP4V-ES	Create Automatically	Edit	Delete
2	camera2	start	virtual	unicast	no	yes	192.168.1.17	MP4V-ES	Create Automatically	Edit	Delete
3	camera3	start	virtual	unicast	no	yes	192.168.1.17	MP4V-ES	Create Automatically	Edit	Delete
4	camera4	start	virtual	unicast	no	yes	192.168.1.17	MP4V-ES	Create Automatically	Edit	Delete
5	camera5	start	virtual	unicast	no	yes	192.168.1.17	MP4V-ES	Create Automatically	Edit	Delete
6	camera6	start	virtual	unicast	no	yes	192.168.1.17	MP4V-ES	Create Automatically	Edit	Delete
7	--	--	--	--	--	--	--	--	--	--	

Figure 4.10 Camera configuration

For using the IPAD/IPHONE/IPOD:

1. Go to Settings and select the WiFi of the reference platform.
2. Go to "**Streamer**" application and select the address of the visual camera you want to view.

3. Go to Safari and type in the IP address taped on the live camera. This live camera has an IP address **192.168.1.x**.[4]

4.5.1 Troubleshooting NVR

1. If a wireless camera is not connecting to the reference platform, you may need to flash the camera.
   (a) Connect the camera to a PC via Ethernet.
   (b) Open Axis IP Utility on the PC and refresh the list until the camera is found.
   (c) Make sure that the reference platform is powered on and broadcasting its SSID. Copy the camera's IP address into a web browser (username: root, password: root).
   (d) Select Setup → Wireless and choose the reference platform's SSID from the list.
   (e) Save the settings.
2. If a wired camera is not connecting to the reference platform through an Ethernet switch, you may need to add the Ethernet interface to the platform's bridge. Refer to the troubleshooting part of Section IV Setup.

4.6 Internet 3G Broadband Gateway

This section describes the procedure to connect the i.Mx6 board to the internet via 3G.[5]

1. Once all the scripts are set up for the specific carrier, please open a browser and type in 192.168.1.1.
2. Navigate to the system tab and make sure that the board is functioning in router mode. If not, select it and apply changes.
3. In the terminal, type in: pppd call gprsdial & (to run this command in Background mode).
4. The board should try to connect to the internet via the 3G dongle. It may take some time. You can observe its connection by seeing if the LED on the physical dongle goes a solid color.
5. To test if it made the connection type in terminal: ifconfig ppp0.
6. If the interface ppp0 shows up in the console, then your dongle is connected.[6]

---

[4] If you do not know the IP addresses, then go to Network Video Recorder → Camera to view the IP address in "Camera IP" column.

[5] We are using the Huawei E160 O2 dongle with an AT&T SIM card. To test the SIM card on a laptop, see Appendix D. If you use a different carrier, you will need to edit the scripts on the board under/etc/ppp/gprs-connect for your carrier connections (see http://zoom.custhelp.com/app/answers/detail/a_id/1030) and/etc/ppp/peers/gprsdial for your USB port and dongle type (Figure 4.11).

[6] The Huawei E160 O2 dongle operates under GSM frequency bands: 850, 900, 1800, and 1900, and UMTS frequency band: 2100. If your carrier does not comply with those frequency bands, you will not be able to connect. Likewise, if you use a different dongle, make sure that the carrier and the frequency bands of the dongle match! A good rule of thumb is to check if you can connect via other devices first.

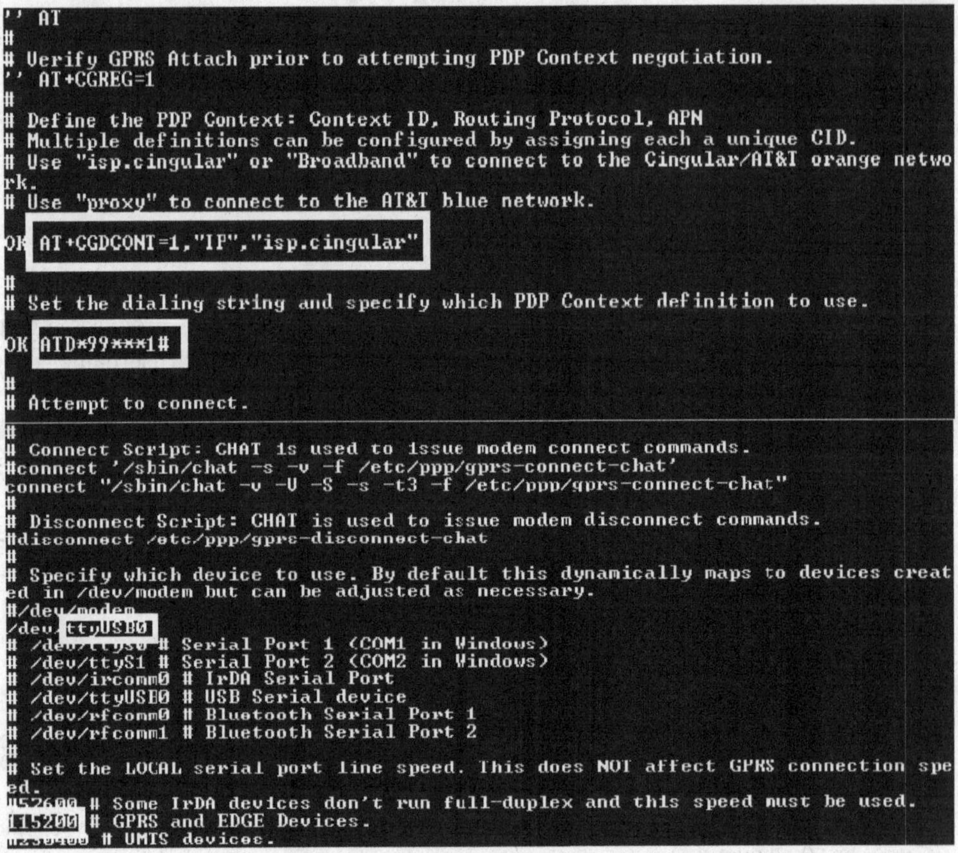

```
'' AT
#
# Verify GPRS Attach prior to attempting PDP Context negotiation.
'' AT+CGREG=1
#
# Define the PDP Context: Context ID, Routing Protocol, APN
# Multiple definitions can be configured by assigning each a unique CID.
# Use "isp.cingular" or "Broadband" to connect to the Cingular/AT&T orange netwo
rk.
# Use "proxy" to connect to the AT&T blue network.
OK AT+CGDCONT=1,"IP","isp.cingular"
#
# Set the dialing string and specify which PDP Context definition to use.
OK ATD*99***1#
#
# Attempt to connect.
#
# Connect Script: CHAT is used to issue modem connect commands.
#connect '/sbin/chat -s -v -f /etc/ppp/gprs-connect-chat'
connect "/sbin/chat -v -V -S -s -t3 -f /etc/ppp/gprs-connect-chat"
#
# Disconnect Script: CHAT is used to issue modem disconnect commands.
#disconnect /etc/ppp/gprs-disconnect-chat
#
# Specify which device to use. By default this dynamically maps to devices creat
ed in /dev/modem but can be adjusted as necessary.
#/dev/modem
/dev/ttyUSB0
# /dev/ttyS0 # Serial Port 1 (COM1 in Windows)
# /dev/ttyS1 # Serial Port 2 (COM2 in Windows)
# /dev/ircomm0 # IrDA Serial Port
# /dev/ttyUSB0 # USB Serial device
# /dev/rfcomm0 # Bluetooth Serial Port 1
# /dev/rfcomm1 # Bluetooth Serial Port 2
#
# Set the LOCAL serial port line speed. This does NOT affect GPRS connection spe
ed.
#57600 # Some IrDA devices don't run full-duplex and this speed must be used.
115200 # GPRS and EDGE Devices.
#230400 # UMTS devices.
```

Figure 4.11 Setup for 3G service provider connections

4.7 UPNP

The following are the steps to enable UPNP features:

1. Enable UPNP features from web GUI.

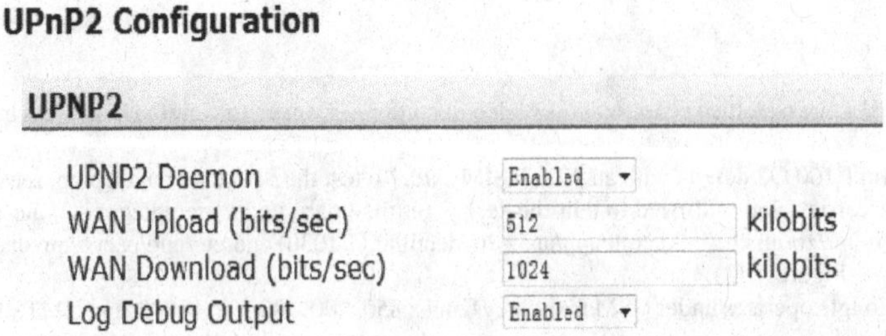

UPnP2 Configuration

UPNP2

UPNP2 Daemon	Enabled	
WAN Upload (bits/sec)	512	kilobits
WAN Download (bits/sec)	1024	kilobits
Log Debug Output	Enabled	

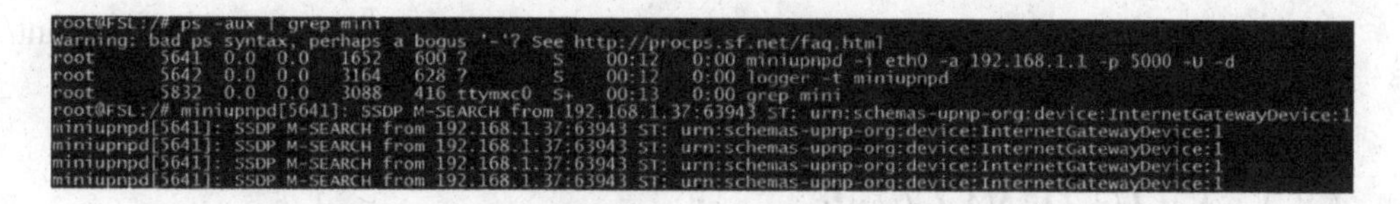

Figure 4.12 UPnP2

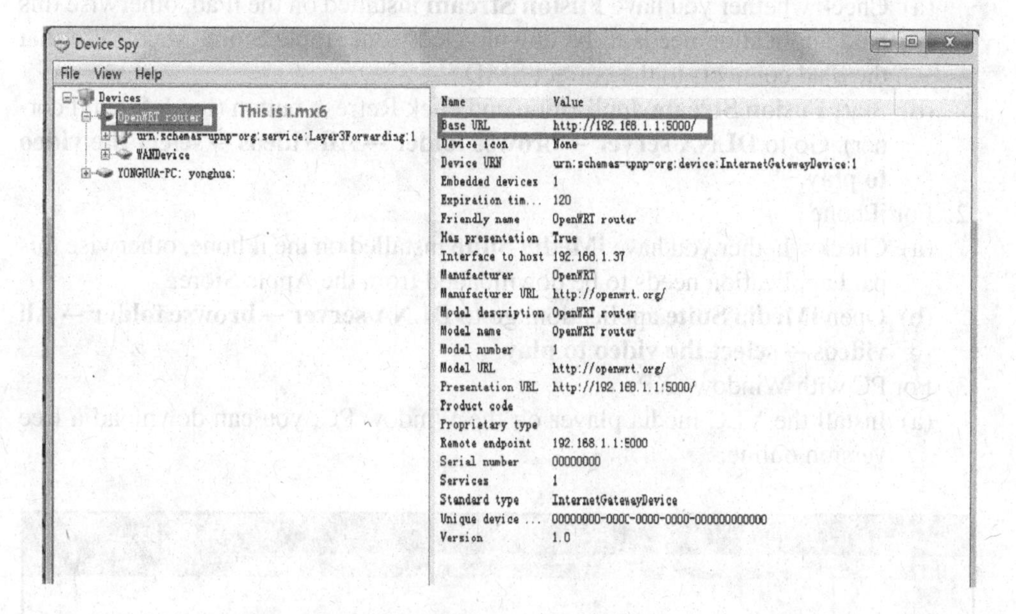

Figure 4.13 uPnP setup

2. At this moment, you can confirm from command line (Figure 4.12).
3. Download a UPNP tools. Such as Intel UPNP developer tools (download from here: http://opentools.homeip.net/dev-tools-for-upnp).
4. Install the UPNP tools on a PC, which connect i.mx6 board by any network (for example, WIFI).
5. Run device spy from the UPNP tools, see Figure 4.13.

4.8 Digital Living Network Alliance (DLNA) Media Server

This section describes the steps required to demonstrate the audio/video streaming capability, using DLNA application, of the Reference Platform.

4.8.1 Set Up Reference Platform as DLNA Server

DLNA server has been pre-installed on the reference platform and the SD card should have a partition with media files. The DLNA server will start when the board is booted and the reference platform will be ready to stream media files to any DLNA compliant clients. Please follow the instructions below to set up the DLNA clients.

4.8.2 Set Up DLNA Clients

1. For iPad
   (a) Check whether you have **Fusion Stream** installed on the iPad, otherwise this paid application needs to be downloaded from Apple Store. Make sure that the iPad connects to the correct SSID.
   (b) Start **Fusion Stream** application and click Refresh button (top left hand corner). Go to **DLNA server → browse folder → All videos → select the video to play**.
2. For iPhone
   (a) Check whether you have **iMedia Suite** installed on the iPhone, otherwise this paid application needs to be downloaded from the Apple Store.
   (b) Open **iMedia Suite** application; go to **DLNA server → browse folder → All videos → select the video to play**.
3. For PC with Windows XP
   (a) Install the VLC media player on the Window PC; you can download a free version online.

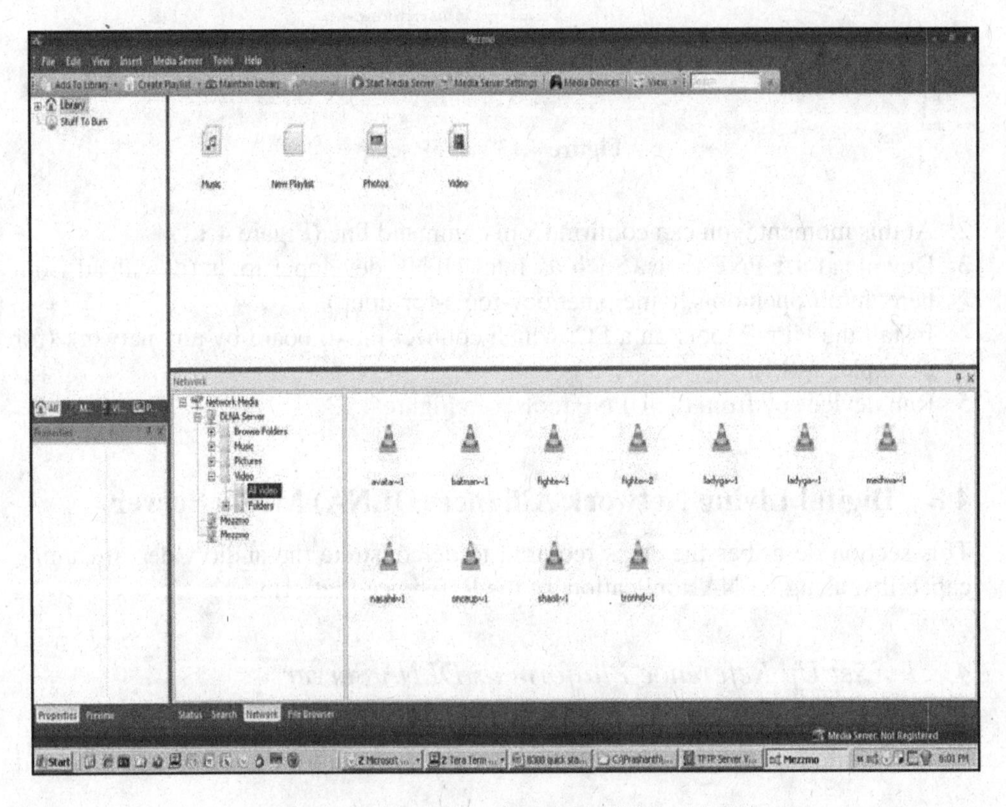

Figure 4.14 Mezzmo DLNA client on PC

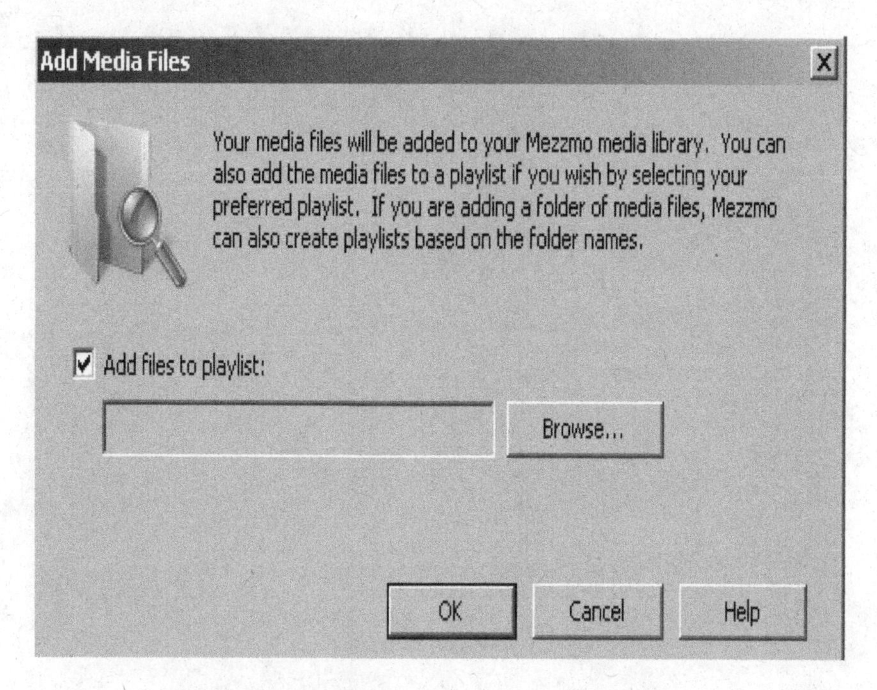

Figure 4.15 Adding media files to DLNA server

(b) Install **Mezzmo** using Google search.
(c) Open Mezzmo
  (i) Click the **Network** tab (on the bottom bar) and select **Network Media/DLNA server/Video/All videos** (Figure 4.14).
  (ii) Click **create playlist** and select the newly created playlist.
  (iii) Click on one of the video files and check **Add files** to playlist. Click **OK** (Figure 4.15).
  (iv) Select the playlist that you want to use. For example: **library → new playlist**.
  (v) Right click on the video file you want to play and select **play**.
  (vi) The video file will automatically open in the browser.

4. For PC with Windows 7
   (a) Windows Media Player 12 can also be used as a DLNA client.
   (b) In Windows Media Player 12, select DLNA server/Videos/All Videos.[7]

References

[1] Utilite (2014). Utilite i.Mx6 Board from Computlab, http://utilite-computer.com/web/home (accessed 10 December 2014).

[2] Freescale (2012). Freescale MCIMX6Q-SDB Sabre Board, http://www.freescale.com/webapp/sps/site/prod_summary.jsp?code=RDIMX6SABREBRD&fsrch=1&sr=3&pageNum=1 (accessed 10 December 2014).

[7] To check the CPU utilization and the processes those are running at any time: Go to **System → Diagnostics** and type **top** in the dialog box in front of the **Command Run** and click **Command Run**.

(v) The video file will automatically open in the browser.

4. For PC with Windows 7:

References

5

C-IoT Cloud-Based Services and C-IoT User Device Diversity

5.1 C-IoT Cloud-Based Services

5.1.1 Introduction and Drivers to C-IoT Service Platform

Internet of Things (IoT) vision is expected to be distributed cloud-based computing, advanced sensors and big-data analysis.

5.1.1.1 RFID and Sensor Network

In the IoT paradigm, many of the objects that surround us will be on the network in one form or another. Radio Frequency IDentification (RFID) and sensor network technologies will rise to meet this new challenge, in which information and communication systems are invisibly embedded in the environment around us. These result in the generation of enormous amounts of data, which have to be stored, processed and presented in a seamless, efficient, and easily interpretable form. This model will consist of services that are commodities and delivered in a manner similar to traditional commodities. Cloud computing can provide the virtual infrastructure for such utility computing which integrates monitoring devices, storage devices, analytics tools, visualization platforms, and client delivery. The cost-based model that Cloud computing offers will enable end-to-end service provisioning for businesses and users to access applications on demand from anywhere.

Advancement in technology in the networking and telecom areas drives a new wave of architectural design and new Market Opportunities. More details are provided in [1].

Collaborative Internet of Things (C-IoT): For Future Smart Connected Life and Business, First Edition.
Fawzi Behmann and Kwok Wu.

5.1.1.2 Sensor Fusion

Sensor fusion is a technology that has come of age, and at just the right time to take advantage of developments in sensors, wireless communication, and other technologies. With sensor fusion and sensor hub chips, it is now possible to efficiently interface to a variety of digital sensors.

A current trend combines a microcontroller with three or more MEMS sensors in a single package. The microcontroller is an ultra-low-power that process data sensed by external sensors such as for gyroscope, magnetometer, and pressure sensors. Functioning as a sensor hub, it fuses together all inputs with a set of adaptive prediction and filtering algorithms to make sense of the complex information coming from multiple sensors.

While quite a lot of functionality can be achieved at a local level, interaction with the Cloud is where the fun really begins. Remote sensor data can be processed by a sensor fusion device and sent to the Cloud for recording, further analysis, or even to trigger an action.

5.1.1.3 Social and Sensor Data Fusion in the Cloud

Mobile phones increasingly become multi-sensor devices, accumulating large volumes of data related to our daily lives. At the same time, mobile phones are also serving as a major channel for recording people's activities at social-networking services in the Internet.

These trends obviously raise the potential of collaboratively analyzing sensor and social data in mobile cloud computing. Participatory sensing, for instance, enables to collect people-sensed data via social network services (e.g., Twitter) over the areas where physical sensors are unavailable. Simultaneously, sensor data are capable of offering precise context information, leading to effective analysis of social data.

5.1.1.4 Gateways

Capitalize on strong presence of gateways – connecting to the Embedded local area network (LAN) and wide area network (WAN) – Cloud Computing companies are announcing initiatives to create standards for gateways that can deal with a flood of data from devices associated with the "IoT."

Software may be a bigger piece of the IoT puzzle than connectivity itself. With sensors, cameras, machinery, medical equipment and countless other objects gathering data, and various applications processing and analyzing that information, something will need to tie all those elements together. Smart portable device companies are announcing a cloud-based platform for developing back-end IoT applications, with the aim of saving effort for enterprises.

5.1.1.5 C-IoT Service Platform

For the realization of a complete IoT vision, an efficient, secure, scalable, and market-oriented computing and storage resourcing is essential. Cloud computing is the most recent paradigm to emerge, which promises reliable services, delivered through next generation data centers that are based on virtualized storage technologies. This platform acts as an aggregator of data from multiple ubiquitous sensors; as a computer to analyze and interpret the data; as well as providing the user with easy to understand web-based visualization. The service platform usually consists of sensor fusion software components, which performs ubiquitous sensing and processing works in the background, *hidden* from the user.

The service platform will consolidate data collected from multiple appliances from multiple IoT service providers. This represents a significant milestone in the evolution of the IoT, establishing a standardized and secure platform for service providers to quickly and cost-effectively introduce differentiating IoT services.

Heavy-duty processing could be done inside the service platform, or it could happen inside the web-connected data centers, or the cloud.

Cloud computing promises high reliability, scalability, and autonomy to provide ubiquitous access, dynamic resource discovery required for the next generation IoT applications. Consumers will be able to choose the service level by changing the Quality of Service parameters.

The C-IoT service platform will support applications in smart energy, smart meters, tele-health, and other smart home services.

5.1.2 Classes of C-IoT Cloud Computing

Similar to cloud computing, there are three primary classes of IoT cloud computing: public, private, and hybrid. Cloud computing service providers (CCSPs) such as Google and Amazon that provide services over a WAN are considered to be providing public cloud computing solutions. There are three categories of public cloud computing solutions. They are:

5.1.2.1 Software as a Service (SaaS)

Software as a Service (SaaS) applications have been growing rapidly ever since the mid-2000s, led primarily by the CRM category, and followed by other departments such as HR, finance, and marketing. In fact, the current buzz in the SaaS industry is all about the "marketing cloud."

In one form of SaaS, an independent software vendor (ISV) such as Salesforce.com hosts an application in one or more of their own data centers. In another form of SaaS, an ISV such as Virtual Bridges hosts an application in one or more data centers provided by an Infrastructure as a Service vendor such as Amazon. In either case, the application is provided to users on a usage basis.

5.1.2.2 Infrastructure as a Service (IaaS)

The two primary forms of infrastructure as a service (IaaS) are compute and storage. Providers of IaaS solutions typically implement their solutions on a virtualized infrastructure and charge for them on a usage basis.

5.1.2.3 Platform as a Service (PaaS)

Platform as a service (PaaS) is the delivery of a computing platform and solution stack as a service. PaaS offerings from providers such as Force.com include workflow facilities for application design, application development, testing, deployment, and hosting as well as application services such as web service integration, database integration, and security.

There is significant interest in public cloud computing solutions, most notably SaaS and IaaS. However, due in part to concerns about security and data confidentiality, most IT organizations will, like Bechtel, decide to adopt the same techniques internally as is used by CCSPs such as Google and Amazon. That approach is referred to as private cloud computing.

A hybrid IoT cloud computing solution involves a combination of services provided by the IT organization itself as well as by one or more CCSPs. For example, an IT organization may either already have, or be in the process of acquiring a four-tier application. The IT organization may decide that for security reasons that it wants to host the application and database servers itself. However, in order to improve the interaction with the users of the application, the IT organization may also decide to let a CCSP host the web tier in numerous data centers around the globe.

As noted, private cloud computing involves IT organizations implementing the same techniques themselves as are typically associated with public cloud computing solutions.

5.1.3 C-IoT Innovative and Collaborative Services

Cloud-based services are the future for creative, innovative, and collaborative services. IoT Innovative and Collaborative Services represents the aggregate of IoT Point solution that analyzed in a collaborative fashion and is scalable by the complexity, security, and performance of an application. Here are some potential areas of IoT cloud-based services:

- Leveraging SaaS to drive Innovation and Results. Competition is intense. Market forces are increasingly volatile. And business leaders are under even more pressure to drive innovation. How can you make every area of your organization more

effective – and do it faster than ever, with fewer resources? The answer is in the cloud. How sales, finance, marketing, procurement, and supply chain leaders are using cloud solutions – and SaaS in particular – to become more agile, efficient, and customer-focused. Such solution will respond to: (i) How industry leaders are using SaaS capabilities to drive business process innovation across their value chain. (ii) Ways to balance the needs of line-of-business and IT decision makers to deliver more customer value faster. (iii) How you can reduce time-to-value and respond to market changes faster using SaaS solutions
- Make your Business Smarter with Cloud-based Analytics. Every industry faces incredible growth in the amount of data collected each day, even each hour – customer behavior, market changes, investments, supplier performance, sales figures. Cloud analytics solutions are designed to handle this tidal wave of data while providing insights quickly, delivered through SaaS can create competitive advantage using traditional business intelligence methods and deeper analytics embedded within business processes. SaaS-based business analytics can become more agile, serve the customer better, and deliver business results.
- Saas connect people with collaborative Business Networks. Now more than ever, business gets done through connections. Information sharing, community building, and collaborative planning are necessary ingredients for staying effective. SaaS is rewiring the approaches for how people and information are aligned. Entire value chains and business ecosystems are now collaborating in the cloud in ways that were not possible in the analog world. Internal teams can access information in more fluid, more intuitive ways.
- Saas provides collaborative intelligence by correlate data received from Fitbit with data from Nest thermostat and LED lighting such that when, say going to sleep, with just moving your hand with Fitbit wearable device, it trigger two actions adjusting the temperature of the thermostat and dim the LED light.

5.1.4 The Emerging Data Center LAN

There are four-stage evolutionary path that IT organizations can take to evolve their data center LANs. A critical characteristic of the evolutionary path is that it is flexible. Virtualization plays key role is the consolidation and management of data centers. Data centers consume a lot of power. Energy management starts from the SoC to server. Example on SoC energy management is provided in [2].

5.1.4.1 Distributed Servers

As part of stage 1, most IT organizations begin the process of assessing the cost/benefits of a centralized versus distributed servers. In some cases, consolidating servers out of branch offices and placing them into centralized data centers could result in reduction of cost and enables IT organizations to have better control over the company's data. In addition to consolidating servers, during this stage many companies also reduce the number of data centers they support worldwide and most begin to virtualize at least some of their data center servers. During this stage, some IT organizations begin to implement LAN switches with 10 Gbps interfaces to support the anticipated increase of I/O needs.

The overall approach to data center design at this stage is characterized by having functionality such as servers, storage, LAN switches, firewalls, and load balancers be both manually configured and dedicated to a single service or application. This approach to data center design and management results in an increase in the overall cost of the data center. It also increases the time it takes to deploy a new service or application since new infrastructure must be designed, procured, installed, configured, and tested before a new service or application can go into production.

The trend is to have hybrid clouds interconnecting global clouds with private clouds.

5.1.4.2 Data Center Virtualization

Everything Virtualized – Gateways, network appliances/servers, storage servers, and sensor nodes.

One of the key characteristics of stage 2 is that during this stage IT organizations implement server virtualization more broadly. The increased deployment of server virtualization at this stage leads to greater flexibility and the increased movement in server-to-server communications. Another change that occurs during stage 2 is that as IT organizations deploy servers with an increased number of cores, the number of virtual machines (VMs) per physical server typically increases proportionally. One impact of the increase in the number of VMs per physical server is that the network I/O requirements of the multi-core physical servers that have been virtualized in stage 2 begin to exceed the capacity of the existing GbE and multi-GbE aggregated links.

However, the biggest change to the traditional IT model that occurs during stage 2 is that driven both by the need to save money and to reduce complexity, many IT organizations implement a two-tier data center LAN, consisting of access and core switches, in at least one of their data centers.

5.1.4.3 Cloud Optimized Data Center LAN – Hybrid Clouds

In the traditional data center LAN environment, the data network is kept separate from the storage network. In stage 3, some IT organizations begin to experiment with deploying a unified data center switching fabric.

In stage 3, many IT organizations begin to deploy a new generation of Lossless Ethernet technologies that are based on a collection of standards that are commonly referred to as IEEE Data Center Bridging.

During stage 3 the bandwidth efficiency and availability of Layer 2 data center LANs with redundant links can be greatly improved by assuring that the parallel links from the servers to the access layer and from the access layer to the core layer are always in an active–active forwarding state.

Another major change that occurs in stage 3 is that many IT organizations begin cross train key employees and by creating goals and a reward system that encourages employees to take a more holistic view of IT. They also do this in part by beginning to implement common tools such as service orchestration.

5.1.4.4 The Near Term Future

There is no doubt that data center LANs will continue to evolve past what IT organizations achieved in stage 3. For example, virtually all IT organizations will continue to expand their implementation of a unified fabric within data centers and to implement continuingly more sophisticated automation. Many IT organizations will also increase efforts to reduce the impact of organizational silos.

Another possible change to data center LANs in the near term future that will build on top of what was already achieved is that many IT organizations are likely to extend their unified fabric both between their own data centers as well as between their data centers and one or more data centers provided by a CCSP. One obvious advantage of doing this is that it enables IT organizations to efficiently move workloads between data centers and hence enhances the ability of the IT organization to implement disaster recover/business continuity solutions.

5.2 C-IoT User Device Diversity

5.2.1 Introduction

Portable and wearable computing introduced major shifts in how humans interact with computing devices and information, dramatically reducing the gap between immediate information and the person for whom that information is the most useful. Health and fitness buffs already wear monitors that record their heart rate and the distance they run, coupling that to a PC to analyze the results. Wearable wireless medical devices include accelerometers to warn of falls, and insulin pumps and glucose monitors for diabetics. Each of these devices can connect to a smartphone via Bluetooth, and can issue an alert to their doctor by triggering a call over the cellular network.

The “cloud” will become more intelligent, not just a place to store data – Cloud intelligence will evolve into becoming an active resource in our daily lives, providing analysis and contextual advice. Virtual agents could, for example, design your family’s weekly menu based on everyone’s health profiles, fitness goals, and taste preferences.

Many products and reference designs for wearable devices were released to the market. Examples: smart earbuds, smart headset, and a smartwatch to a device – developed with Rest Devices for Baby product line – that can be worn on an infant’s onesie that monitors the baby’s vitals and sends the data to a coffee mug, where it can be displayed. Devices such as the Fitbit come with iOS and Android apps, include capabilities for social sharing, and track everything from sleeping habits, to the number of steps taken everyday.

With open-source OS like Android, the fast growing IoT and sensor's technology are empowering mobile developers to connect and control objects like thermostat, light fixtures and experiment with an endless range of products.

Access to complete Bluetooth Smart SoC, plus embedded Wi-Fi, enables OEMs of all sizes to seamlessly deliver connectivity to battery-operated devices for sports and fitness, health and wellness, security, automation, and more.

Making available Smart Development Kit through strategic distribution partners will accelerate time-to-market of IoT products and solutions that matter. The hardware development kit opens the door for original equipment manufacturers (OEMs) to develop products for the "IoT" with ease. It provides access to a low-power client device with an integrated Bluetooth Smart (formerly Bluetooth Low Energy) software stack and application profiles, offering OEMs an easy-to-use, cost-effective embedded wireless solution with a small footprint to inspire connectivity in a new range of devices.

5.2.2 *C-IoT Developers/Platform*

Collaborative Internet of Things (C-IoT) is empowering growth of Entrepreneurs and software developers at three areas:

1. Mobile devices for innovative IoT applications
2. Gateway-based applications
3. IoT cloud services.

Much of *today's* IoT development *focuses on technology*. Should I use 2G or LTE, WiFi or mesh networks? Should I embed connectivity directly in my device, or use a gateway? How do I make robust and compact enough devices? How do I make a wearable with the best battery life? Which of the 50+ IoT cloud platforms is the right one for my application? Is my home automation solution secure against hackers? How can I use artificial intelligence to optimize the maintenance schedule of my factory robots?

5.2.2.1 Entrepreneurs and Software Developers for Mobile Devices

The evolution in mobile since 2008 has resulted into creation of community of entrepreneurs and developers for both iOS and Androids.

Android and iOS won because they used fundamentally different business models from those of Nokia/Symbian, Windows Mobile, and Blackberry. That in turn has created an abundance of apps and devices. This entrepreneur-driven demand creates new markets that are several times bigger than the existing ones.

5.2.2.2 Consumers Drive Application Development for Mobile Devices

The demand for iPhones (and later for Android smartphones) did not come from business users. It came from consumers craving the hundreds of thousands of apps available on these devices. The combined value of all those apps is beyond the utility that Microsoft could ever hope to create on Windows Mobile devices.

Platforms thrive by a steady stream of innovation by developers, experimenting day and night in their quest to find new use cases. The solutions to these use cases drive sales of iPhone and Android phones.

WiFi, social networks, smartphones, and tablets all first found grassroots success with consumers and only later started to appeal to enterprises.

The most newsworthy example is the battle brewing between Google and Apple to win control of the consumer market. Through a series of acquisitions, these two giants are locked in a mega-struggle to make their smartphones the central element of a much bigger market – the smart home – including intelligently connected and controlled lighting, heating, and entertainment.

5.2.2.3 C-IoT Developers

Today's forecasters predict fast, attractive growth of the market as it is. They promise tens of billions of devices in the market by the end of the decade, based on the current state of the market, known demand and technology evolution.

If developers and entrepreneurs adopt IoT technology as swiftly as they did Android and iOS, there will be well over 4 million IoT innovators at work by the end of this decade. See Figure 5.1. Think about how much demand they could create for the IoT industry.

As a contrast, VisionMobile forecasts a fast growth of the IoT developer base in the next years, reaching well over 4 million innovators and entrepreneurs by the end of the decade. With every new use for IoT technology that they discover, demand will grow and this market will become more attractive still [3].

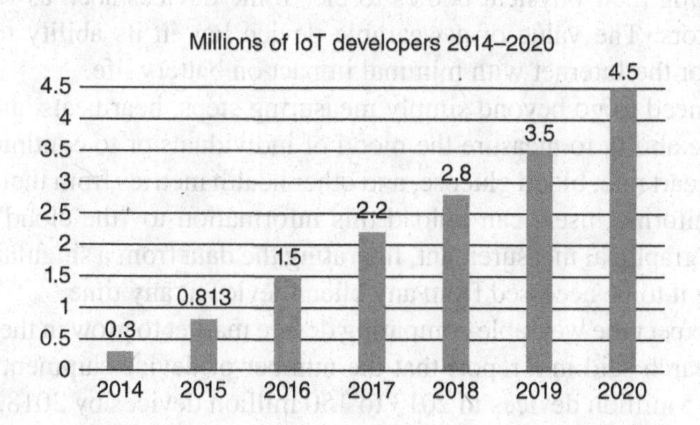

Figure 5.1 IoT developers. *Source*: VisionMobile

Steady stream of developer-driven innovation is already emerging in C-IoT.

- Developers are innovating on asthma inhalers (Propeller Health), basketballs (94fifty), smoke detectors (Nest), and bottles of booze (Beverage Metrics' Tilt). IoT solutions can count pests on the farm (Spensa Technologies), measure radiation around Fukushima (SafeCast), warn you when the sewage system floods (Don't Flush Me), let you express yourself (Switch Embassy's tshirtOS) or protect human rights activists (the Natalia Project).
- Wearables (Pebble, Nike+), connected cars (GM, Ford, Dash Labs), and buildings (SmartThings, Panoptix) are opening up APIs for third-party developers, thereby facilitate development of C-IoT applications that inter-operate among multiple point-solutions of Smart things. This will further drive the wearables market to new heights.
- Hackers are combining IoT devices and APIs into amazing new use cases. Like using a Fitbit activity tracker to pause movies on Netflix when you fall asleep. Or combining Fitbit with SmartThings to lock you out of your house until you have completed your morning run.

These wide-ranging and often unexpected devices, services, and apps that come from a growing community of IoT developers is the main factor that will drive demand for IoT to unseen heights.

5.2.3 Wearable Devices – Individual

Work-life balance itself has so many different dimensions of measurement, that it would be hard-pressed to imagine anyone spending an inordinate amount of time documenting these dimensions manually on their smartphones at frequent intervals. This is where the role of wearable comes in.

Thus, wearable devices are typically associated with consumers. Consumers are now connecting their physical bodies to electronic devices such as Fitbit and other health monitors. The value of a wearable device lies in its ability to connect to a smartphone or the Internet with minimal impact on battery life.

Wearable need to go beyond simply measuring steps, heartbeats, and sleep cycles and attain the ability to measure the mood of individuals or to continuously monitor and control heart rate, blood glucose, and other health metrics from their smartphones. Beyond monitoring, users can upload this information to "the cloud" for real-time analysis and graphical measurement, liberating the data from a singular sensor device and allowing it to be accessed from any client device at any time.

Analysts expect the wearable computing device market to grow in the coming years. Juniper Research said in a report that the number of device shipments will increase from about 15 million devices in 2013 to 150 million devices by 2018.

As the market continues to gain momentum, there will be wide variety of business opportunities in this growing space to offer the breadth of IP and customized components that enable creative new smart wearable devices to be connected.

Today, a handful of companies produce wearable devices to detect brainwaves that infer how calm or attentive a person may be. With further growth in this area, sensors that are powerful enough to stream brainwave signals in real time will be developed. The form factors of these devices would also be compact enough so as to be inconspicuous.

Embedded Devices, SoC integration versus SIP (RCP) versus modules – Industry Applications

As price of sensors continue to decline and microcontroller unit (MCU) processing, the future sensors become more intelligent by providing local processing and limited storage. This becomes attractive to many of the Industry Applications such as energy management. Hence, various design options are examined such as integration of the MCU with the sensor as an integrated SoC, exploring option of System In Package option and having modular approach. Power management is a key consideration in determining the right approach.

5.2.4 Harvesting (Self-Powered Nodes) – Infrastructure Applications

In general, sensors associated with public infrastructure applications such as for asset tracking and logistics may rely on self-powered to eliminate the cost to cover services for battery replacement.

Smart appliance sales are rising. Wearable fitness trackers are creating a new market in quantified health. Apple is making an entry into the home automation and health industries, through IoT applications.

5.2.5 Embedded Devices and Servers

Healthcare and home-automation have been pushing the market forward, but there are plenty of other industries that wearable technology and embedded systems can disrupt. On a macro-level, there are two major entry points into IoT across industries – servers and embedded processors.

Powerful, embedded systems are essential to creating disruptive IoT devices. A lot can be done with low-cost sensors connected to WiFi, but it's advances in high-end embedded devices that will grow the market. This will include healthcare technology, digital signage, industrial automation, and even casino gaming.

As the high-end wearable and embedded market continues to grow, there will be a corresponding growth in the amount of data produced. In fact, a lot of the value of such technology will come from the highly targeted, highly specific data it produces. This means companies will need enhanced data processing capabilities, along with business intelligence software, to sort and analyze the information influx.

In short, more wearable and embedded devices will create more data, which will increase the need for larger data centers. More data centers mean more servers, which equals a huge opportunity for both companies.

In this field, Intel x86-based servers have been the industry standard for years. Gartner estimates that Intel sales are 92% of all server processors in 2013 versus AMD of 7%. Intel also recently invested $740 million in Cloud era, a big data analytics company. Their intention is not to compete with the cloud computing solutions from companies such as Microsoft and Amazon Web Services, but rather to spur the adoption of Hadoop among enterprises. Similar to embedded systems, increased cloud usage ultimately means increased server sales.

AMD, on the other hands, announced a new "ambidextrous" server chip that bridges ARM and x86 processors, a product they hope will appeal to companies looking for greater flexibility. AMD also continues to invest in the low-cost, micro-server market, where they hope their graphics-experience and APU tech will give them an edge. Theoretically, micro-servers are poised for wide adoption in data centers, and could help improve energy efficiency. So far however, there has been little movement, for a variety of reasons.

IoT will continue to change how industries operate, and there are significant opportunities both in embedded systems and servers. AMD seems to have an edge into the embedded world, while Intel is poised to maintain its dominant server position. Intel's resources make them a large threat, especially in the high-end embedded market and in wearable, if they can promote adoption of their Edison processor. Given AMD's projections that the embedded market will soon be worth over $9 billion, it is likely that neither company is going to cede the area without a fight.

5.2.6 Performing Sentiment Analysis Using Big Data

With the human brain containing on average 86 billion neurons, there are potentially trillions of brainwave signals from these wearable devices that need to be analyzed. The amount of time employees stay attentive at work can signify how engaged they are. A sustained series of spikes in brain activity could indicate stressful working conditions. Extreme brain focus at night, followed by lack of restful sleep may imply an organization with a workaholic culture.

Big Data algorithms will be able to correlate these data with several other bodily measurements such as sleep, physical activity, and heart rate to reflect the average sentiment within an organization in real-time. Individuals could monitor their own sentiment privately within their smartphone app, and the general public could view anonymous aggregated data about an organization. Millenials, who already share data about their physical exercises through a combination of wearable devices and social media, will continue to do so with these highly advanced brain-sensing wearable.

5.2.7 *IBM Watson for Cognitive Innovations*

Among the major initiative by companies is that of IBM by establishing early 2014 IBM Watson Group, a new business unit dedicated to the development and commercialization of cloud-delivered cognitive innovations. The move signifies a strategic shift by IBM to accelerate into the marketplace a new class of software, services and apps that think, improve by learning, and discover answers and insights to complex questions from massive amounts of Big Data.

IBM will invest more than $1 billion into the Watson Group, focusing on development and research and bringing cloud-delivered cognitive applications and services to market. This will include $100 million available for venture investments to support IBM's recently launched ecosystem of start-ups and businesses that are building a new class of cognitive apps powered by Watson, in the IBM Watson Developers Cloud. A high level appreciation of Watson winning Jeopardy in 2011 is provided in [4].

5.2.8 *Far-Reaching Consequences*

When these brain-sensing wearable become as pervasive as Fitbits, corporate cultures around the world will be visible for all to see. Employees seeking better opportunities would have an in-depth view into the working dynamics of any particular company. High-school students considering careers in a particular field will be able to evaluate its pros and cons more accurately. With aggregate employee sentiment being widely available, HR departments will become increasingly important in ensuring employee retention.

HCM software is the closest thing to measuring and tracking information about employees. However, none of the existing cloud HCM solutions adequately capture the sentiment of their most valuable assets – their employees.

During the post World-War II economy, manufacturing equipments that could produce a better widget faster and cheaper were regarded highly as prized assets because they were an important part in providing a competitive edge. In the knowledge economy, happy, engaged, and intellectually stimulated employees are the ones that provide their companies with competitive differentiation. That is why the next cloud revolution, the People Cloud will not be about employers using HCM software to measure the numerical worth of their employees, but rather millions of workers driving change in their organizations through real-time sentiment data.

The world around us is changing at speeds never seen before. Mobile devices are collecting countless levels of data through innovative technologies such as motion coprocessors, gyrometers, A, ambient light sensors, application analytics, and more.

We are also seeing a huge increase in the number of "smart" homes equipped with connected lighting, air conditioning, all the way to internet-connected refrigerators. All this is on top of the numerous connected devices already found in modern homes such as iPhones, iPads, printers, connected TVs, game consoles, and so on, and did I mention the rise in connected cars?

With the huge leaps that digital technology has made over the past few years, we are starting to see companies tapping deeper and deeper into these massive data streams, positioning us on the verge of a new era of analytics and measurements.

5.2.9 *C-IoT (Collaborative IoT)*

OEMs creating wearable products require interoperable technology that will allow these new devices to connect with smartphones, tablets, and wireless gateways available today. Since many companies are able to power the Wi-Fi and Bluetooth in the majority of smartphones on the market today, the wireless SoCs are an ideal choice for OEMs developing consumer products that are designed to seamlessly communicate with other mobile devices on the market. The C-IoT service platforms leverage sensor fusion software framework components, to perform ubiquitous sensing and processing works transparent from the user. The will enable the development of C-IoT applications that interoperate among multiple point-solutions of Smart things. This will further drive the IoT market products to new heights.

References

[1] Behmann, F. (2010) Technical trends in telecom/datacom – drive new wave of architectural design and new market opportunity. IEEE Consulting Networks, Austin, TX, November 17, 2010, http://ewh.ieee.org/r5/central_texas/cn/presentations/TelNet%20-%20Presentation%2011-17-2010%20Fawzi%20Behmann%20Final.pdf (accessed 18 November 2014).

[2] Behmann, F. (2010) Energy Management in Power Architecture, Embedded Insights, December 21, 2010, http://www.embeddedinsights.com/channels/author/fawzi-behmann/ (accessed 18 November 2014).

[3] VisionMobile http://www.visionmobile.com/blog/2014/06/who-will-be-the-ios-and-android-of-iot/ (accessed 18 November 2014).

[4] Behmann, F. (2011) The Inner Workings of IBM's Watson. A video interview by Bill Wong of Electronic Design Magazine (Oct. 6, 2011), http://www.engineeringtv.com/video/The-Inner-Workings-of-IBMs-Wats;ESC-Boston-2011-Videos (accessed 18 November 2014).

6

Impact of C-IoT and Tips

6.1 Impact on Business Process Productivity and Smart of Digital Life

Today, Internet of Things (IoT) is affecting our day-to-day interaction with "things" around us and opens the door of possibilities for a more sustainable work environment. In the future, we can expect Collaborative Internet of Things (C-IoT) to generate entirely new job roles and titles and to completely change the way we commute, communicate, and collaborate. How the C-IoT will impact the future of work is divided into the three domains: Individual, Industry, and Infrastructure, which are discussed in the following sections.

6.1.1 *Individual*

- *New Job Roles*. The digital age has ushered in new IT jobs moving from being concerned about installation broadband connectivity to installation and support of complex systems. With the rise of IoT, advance sensing, connectivity, and cloud "big data/analytics" jobs are becoming more specialized than ever. Gartner last year reported that the number of Chief Digital Officers (CDOs) is on the rise, predicting that by 2015, 25% of companies will have one managing their digital goals. The Data Scientist too has become an important asset for companies embracing the value of big data and analytics, and we will begin to see more chief data scientists, analysts, and even chief customer satisfaction officers; moreover, probably some titles we cannot even imagine yet.
- *Productivity at Work*. The rise of social has given way to a new age of communications and team collaboration. Value tools such as Box, Skype, and even Facebook have captured the attention of the next-generation workforce. Video collaboration and imaging will take hold as millennials and digital natives rely on text messaging,

Collaborative Internet of Things (C-IoT): For Future Smart Connected Life and Business, First Edition.
Fawzi Behmann and Kwok Wu.

FaceTime, and even "Hangouts" for true integrative communication at work, saving time, and blurring social tools with modern collaborative work systems.

- *Smarter Water Cooler Chat.* Even water coolers can be connected in IoT, making a trip to the water cooler smarter than ever. The water cooler (coffee machine, etc.) can remember personal preferences, be voice and motion activated, and even deliver drinks on demand without the wait (cutting down on the proverbial water cooler chatter).
- *Plan Workdays around the Weather.* With a more virtual workforce and flexible workdays, weather can impact team productivity and commute decisions. Today, weather forecasts rely on a few satellites or ground-based weather stations as the primary data gathering points from the sky. In the future, billions of sensors will be integrated in different "listening" devices and stations – both in the sky and on the ground. Using Big Data to better predict the Earth's "heart beat" will enable more sophisticated and accurate weather and climate change predictions. For commuters, this will mean more accurately predicting rain, sleet, and snow, well in advance, so that we can better plan our workweek, the days we head into the office (and conversely, the nicest days to stay at home) and even how we commute (Rain today? Think I will take the train.).

6.1.2 Industry

- *Physicians at Work.* The IoT is going to change the way a physician works too, as well as the patient experience at the physician's office or in the hospital, and the overall physician–patient relationship. Today, a patient's condition must still be assessed live in the presence of the physician, face to face. In the future, IoT will enable devices to read data directly from a person's body, enabling physicians to access real-time patient data remotely. New technology also means that the physician can meet with the patient from halfway around the world, changing the shape of where and how their workday happens.
- Medical applications are about to undergo massive growth as it becomes increasingly possible to monitor patients with tissue-implantable wireless-linked sensors. This is one of the most important of the emerging wireless technology trends as these applications can potentially improve the quality of life for so many who suffer from a chronic medical condition. For example, a patient with a condition such as diabetes could be monitored continuously. A small implanted sensor could send blood glucose information to a base station via a wireless connection. If the glucose level moved outside a preprogrammed range, the base station would send an alarm to the appropriate caregiver.
- *Imagine a more Efficient Commute.* About 15% of commute time is spent in traffic, and about 17% of fuel wasted in cities by drivers sitting at red lights. Sensors on our roads, traffic video cameras, and median divides will impact how our vehicles will "talk" to drivers. By monitoring traffic speed, stoplights, accidents, and current road conditions, programmable cars and even roads will push the most efficient routes

to drivers' mobile devices, cutting down commute time, saving gas money, and ultimately making our roads safer.

- *Proactive Customer Services.* After a product is shipped, the interaction between the customer and the vendor usually subsides, at least until the next buying period or a problem arises. Proactive technology can keep a "pulse" on the health of products and "things" to pinpoint issues before they arise. In this era of next-generation customer service, proactive product monitoring means a company can keep customers happy, watch product health around the clock, and avoid any problems.
- *Give Structure to Unstructured Data.* "Big Data" is not just big … it is huge. If leveraged well, Big Data can create new value across the business when unstructured data is converted into structured data. Analyzing data and breaking it down into meaningful intelligence and analytics can tell a richer story about customers, product behavior, market position, employee productivity, and even predicted future success.
- *A "Greener" Business.* Sensory meters have already seen "light" in a few office buildings and homes today, but this will become a necessity in building standards for modern building infrastructures. Installed movement sensors can turn off/on lighting fixtures, heaters/ac, coffee machines, and even the TV as humans move throughout the space, or go home for the night. These sensors are already integrated into blinds, basing temperature and sunlight for how far they open and close, which can improve energy efficiency and production, saving money and the environment.
- *Location, Location, Location.* IoT will make location tracking simpler. Currently done via phones, cars and even in hospitals, Internet-connected equipment and devices will be geographically tagged, saving valuable resources such as time and money. Companies will be able to track every aspect of their business, from inventory to fulfilling orders as quickly as possible to locating and deploying field services and staff. Tools, factories, and vehicles will all be connected by location-based technology making the entire chain ever more efficient.
- *Operational Efficiencies.* Data you collect from your factory floor, logistics network, and supply chain can reduce inventory, downtime due to maintenance, and time to market. You can also use that data to simplify operations.

6.1.3 Infrastructure

- *Improved Safety and Security.* Sensors and video cameras can help monitor equipment to improve workplace safety and guard against physical threats. Connected incident response can coordinate multiple teams to resolve situations faster.
- *Distributed Intelligence and Control.* More frequent, remote software upgrades and enhancements can extend the efficiency and value of your resources, products, and services.
- *Faster and Better Decision Making.* Distributing intelligence and control offloads repetitive decisions and can help prioritize decisions that people need to make.

- *New Business Opportunities and Revenue Streams*. More and new ways to analyze data can help identify new potential markets and business opportunities.
- Social networking is set to undergo another transformation with billions of interconnected objects. An interesting development will be using a Twitter-like concept where individual things in the house can periodically tweet creating Tweet of Things (TweetOT). Although this provides a common framework using cloud for information access, a new security paradigm will be required for this to be fully realized.

6.2 Considerations of Developing Differentiated C-IoT Solutions

The IoT can potentially transform nearly every industry and change the way we live and work, locally and globally. Companies in all industries face challenges to build infrastructures that meet the changing requirements of scale and data management, run on standards, and are highly secure and interoperable. IoT partner ecosystem will ensure that client's migration and connectivity management are done effectively and more securely.

Following is a set of key considerations in developing differentiated C-IoT solutions, which will have an impact on improving business processes and on the quality of our lives and improving business processes.

6.2.1 *Software Processes and Platform*

Over the last decade, emphasis for product differentiation has been on software. According to some estimates, more than 60% of total Embedded Development resources are spent on software. Hence, many companies have put in place software strategy to help their customers to reduce the total cost of ownership of embedded systems.

To this end, many companies have developed process to encompass the entire embedded software lifecycle management – prototyping, development, and testing – powered by simulated development environment representative of the "target hardware." This enables a development of a platform with application that can be developed on a wide range of MCU/MPU (micro-controller unit/microprocessor unit) devices, based on customers' target cost/performance constraints.

6.2.2 *Standardization*

National and international standards provide a common language not just for physical products and equipment, but they are increasingly important for describing processes, procedures, skills, and metrics. Roadmap participants identified a variety of capabilities that must be established through the development, promotion, and adoption of standards to enable safe, efficient, and interoperable hardware and software systems.

Importance of Standardization can be exemplified in the following two case studies: Containerization and Traceability of order.

6.2.2.1 Containerization

Global-standard, smart, modular, designed-for-logistics containers (replacing current cartons, boxes, and pallets) are needed that can accommodate raw materials and finished goods. These include unit-load, carton, and transportation containers.

These containers should be reusable and reconfigurable, providing the ability to take them apart and combine components to support a variety of sizes and shapes while supporting efficient automated handling.

By 2025, major intermodal hubs throughout the United States should have the ability to handle standardized containers at the unit-load and carton level, plus load/unload integration with freight containers.

6.2.2.2 Cloud-Based Visibility and Traceability of Order

The lack of standards for interoperability of the various applications used in the supply chain causes inefficiencies within and across supply chains. Standardization to support collaboration is needed to provide plug-and-play capability between trade partners. End customers and consumers are increasingly driven by visibility-based decisions that reduce variability and time to deliver.

Great strides have been made to develop transaction standards such as those in electronic data interchange (EDI), but this capability is only in its infancy compared to the standardized interoperability that will be required going forward. Standardized modular interfaces between software systems should support collaboration throughout increasingly virtual organizations. Virtual product models should be easily accessible and modified in the cloud by all pertinent players within a supply chain.

As corporate data systems shift from local physical processing and storage to cloud-based systems, it becomes far easier to develop standardized interfaces between systems of trading partners. Highly customized or in-house developed systems are becoming less common. Instead, systems that provide modular, standardized interfaces to provide for interoperability with trading partners will be more common.

By 2025, most applications accessed by logistics and supply chain professionals should be cloud based and standards compliant.

6.2.3 Sensors and C-IoT

Advances in sensor technology continue, making at least part of the IoT all but inevitable by 2025. To fully realize the vision, sensors must be made smaller and more powerful, allowing them to broadcast farther and in difficult environments such as transportation containers and packages.

Sensors at this scale will create a torrent of raw data that must be converted into useful information. The data will be fully useful only if there are standard, universally defined formats for sensor output. Data from sensors as well as data and information from software packages must be exchanged using common protocols, and not common software platforms, and that data must be shared securely.

6.2.3.1 Standardized Ways of Handling Data from Sensors

A common complaint in the material handling and logistics industry is the need to work with data in different formats. Lack of data format standards requires post-processing for analysis or comparison with other data, and this sometimes requires human intervention. The vision for 2025 includes real-time, automatic control of many pedestrian supply chain decisions (i.e., automating the mundane), which cannot happen without real-time data in standard formats from participating sensors.

By 2025, universally accepted standard data formats for all types of sensors should be established.

6.2.4 Advertising Ecosystem Value Exchange

Today's digital native teens understand the value exchange of the advertising ecosystem and are willing to share some information about them in order to get relevant content and offers. They are not afraid to share data, and they opt into everything. They understand the model and are actively participating.

It is time for advertisers to start leveraging all that data to see what Analytics 3.0 will enable us to do and come up with innovative ways to use it to reach audiences more effectively.

Devices will soon offer storage and analytical capabilities that will go way beyond what we are dealing with today. Just think of the data on your mobile device. Today, we look at browsing history, great. But what if we combine the browsing history with the location data? That will give us a true story of what was consumed and where.

What if we could evaluate device movement once a page is completed loading? This could give us insights on sharing in real time and space (if I see something interesting, I might hand the phone to my colleagues or family to look at it as well, which will create a unique and measurable motion).

6.2.5 Opportunity with Industry Supply Chain for Material Handling

6.2.5.1 Supply Chain Material Handling

There is a great market opportunities for growth in the coming years.

We live in a highly connected world that is complex and becoming increasingly so. In the midst of this complexity, all the pieces must fit and work together to accommodate continuous and sometimes mind-boggling change.

This is the environment for material handling and logistics in 2014. Material handling and logistics provide the connections that move goods through the supply chain and into consumers' hands. The impact of the industry on the U.S. economy is extremely broad, touching everything from raw materials at the point of origin to final delivery at the front door of the consumer to recycling and end-of-life disposal.

Trends such as e-commerce and relentless competition are well underway and moving toward maturity. Others, such as Big Data and the IoTs, are in the early stages of development. Each trend has the potential to have a tremendous influence on the material handling and logistics industry in the future.

Sensors, data, and algorithms: imagine a world in which physical objects are able to communicate with people and information systems with low-cost sensors. Imagine a world in which nearly every fact a company needs is available instantaneously. Imagine a world in which sophisticated algorithms make low- and mid-level decisions optimally and automatically, leaving humans to perform tasks that require judgment and intuition. That world will be here by the year 2025.

Forrester Research estimates that online retail will comprise 10% of all US retail sales by 2017, reaching $370 billion, compared to 8% compound annual growth rate (CAGR) in 2012 and 2013. The report cites two underlying causes: (i) increasing use of mobile devices, leading consumers to spend more time online and (ii) traditional retailers making greater investments in e-commerce fulfillment and Omni-channel distribution systems. Which is to say, demand and supply are working together to increase the size of the market.

As consumers become more experienced buying online, Forrester says, they typically move from buying relatively small and inexpensive items such as music CDs and books to pricey, more involved purchases, such as furniture and appliances [1].

Order fulfillment for e-commerce is challenging on at least three fronts.

First, delivery directly to consumers requires very fast order fulfillment times. The time to pick, pack, and ship is no longer measured in days or hours but in minutes.

Second, direct-to-consumer order fulfillment involves handling individual items rather than cases or pallets. Such "broken-case" order picking continues to be very labor-intensive and complex.

Third, sophisticated inventory policies are needed to ensure that products are in stock, but without creating excessive (and expensive) safety stock levels.

By 2025, the challenge for the material handling and logistics industry is not only supporting the demands of e-commerce, but providing true, Omni-channel distribution systems to support the wide variety of means through which consumers will demand their products.

By 2025, the material handling and logistics industry must be capable of supporting a highly diverse set of order and distribution channels in keeping with mass customized products and delivery methods. Customers will want to order with their phones, mobile devices, and computers, as well as through traditional retail outlets, kiosks, and perhaps as-yet-unimagined channels. Delivery modes will be just as diverse from time-definite, long-lead-time delivery to next-day delivery, same-day delivery, and even same-hour delivery.

6.2.5.2 Total Supply Chain Visibility

Although systems already exist for customers to track shipments (FedEx, UPS, etc.), the level of detail demanded in the future will be much greater. For example, consumers can already know that a shipment was last located in a transit facility in Phoenix, but not that the shipment is tied up in traffic in Dallas or that it was placed on an earlier flight in Chicago. For commercial shipments between companies, even more precise information will be needed.

Precise location of services could facilitate new modes of delivery to consumers, including delivery directly to an individual rather than to an address. In such settings, consumers would want to know the precise location of delivery drivers, in addition to drivers needing to know the location of the customer.

For companies in the supply chain, visibility is about more than knowing the location of an item. Available data should include current and historical environmental conditions such as temperature, humidity, and exposure to vibration. Other data could also be collected depending on the application.

By 2025, all shipments should be traceable in real time from the instant of order to the instant of delivery, in transit and in facilities, at the level of individual items and independent of carrier and transportation mode.

6.2.5.3 Deployment of GPS Capabilities across Transportation Assets

Essential to accomplishing universal, real-time tracking is the ability to know the precise location of transportation and other material handling assets. Although global positioning system (GPS) technology is currently available, its deployment is still not universal across transportation assets.

By 2025, all transportation assets should be trackable by GPS.

Another need is real-time locator systems (RTLSs), which track and communicate locations of items to tracking systems within a warehouse, for example. Such systems are in their infancy in 2014; by 2025, they should be widely used.

By 2025, RTLSs should be integrated into total supply chain visibility systems for access by supply chain partners and consumers.

6.2.5.4 Development of Arrival Time Estimation Methods

Although knowing where a shipment is located is important, ultimately customers want to know when a shipment will arrive. Needed are techniques to estimate remaining time for delivery. Such methods should account for distance, road conditions, traffic, and ultimately should be able to predict time of arrival within 1% of remaining time. By that standard, the arrival time of a shipment with 24 h remaining could be predicted within about 15 min. Arrival time for a shipment 4 h away could be predicted within 2.5 min.

By 2025, arrival time estimation methods should reliably be within 1% of remaining delivery time.

6.2.5.5 Integration of Tracking Capabilities across Carriers and Providers

The very nature of supply chains means tracking information must come from multiple service providers. Coordinating this information will require both technological advances and, more importantly, organizational structures to ensure that the interests of all parties are protected and those validation standards are followed.

By 2025, new protocols should be in place to track individual items throughout their lifecycle, recognizing that individual items might be transformed into other products or shipping units and back again as they make their way along the supply chain.

6.3 Practical Tips on Maintaining Digital Lifestyle

6.3.1 Mobile and Wearable Computing

The Internet has ushered in an information revolution by separating knowledge from its physical manifestation (books) and by allowing worldwide, instantaneous access on personal computers. The most recent advance in the revolution has been to divorce access to knowledge from stationary computing devices. We have arrived at the point in history in which it is possible to acquire knowledge, communicate with others, act on decisions, and engage in commerce at any moment from any location.

Mobile computing is changing the way we live and at a pace that few could have imagined even 10 years ago. In 2006, Steve Jobs announced the iPhone and "the Internet in your pocket." Just 7 years later in 2013, some 56% of American adults own and use a smart phone. A survey in 2013 reported that 37% of American teens (ages 12–17) have a smart phone, up from 23% in 2011. Three quarters of teens in this age group reported being "mobile Internet users" on cell phones, tablets, or other mobile devices.

Use of mobile location-based services is also on the rise. Embedded GPS capability in mobile devices allows users to gain useful information related to their current locations, but it also allows apps, algorithms, and other users to know where they are. Users appear increasingly willing to offer this location information. For example, a Pew Research Center study reports that 30% of all social media users tag posts with their locations and that "74% of adult smart phone owners ages 18 and older say that they use their phone to get directions or other information based on their current location."

The next wave in mobile computing appears to be "wearable computing," in which a computing device or a collection of sensors is embedded in a small, wearable accessory such as eyeglasses, a wristwatch, or even fabric in clothing. Applications are currently used by the health care research community to monitor physiological and environmental data of patients. Google has introduced Google Glass, a small eyeglass-based computer that allows users to access the Internet, record and share audio or video, and interact on social media in real time and hands free. Such devices make possible a life – and workplace – of continuous digital input, sharing, interaction, and recording.

By 2025, Individual, Industry, and Infrastructure must be taking full advantage of mobile computing technologies. For example, constantly connected consumers will demand to know where their shipments are and how much longer they will have to wait for them.

6.3.2 *Robotics and Automation*

Advances in robotics and automation continue at breakneck speed. While the headlines are mostly filled with innovations in personal electronics and mobile computing, significant advances are also being made in technology related directly to material handling and logistics. Participants in the Roadmap workshops identified several areas that will have a major impact on the industry in 2025 including robotics, autonomous control, driverless vehicles, and wearable computing.

The robotics industry is in the midst of a true revolution as capabilities increase and costs decrease. More than 160 000 robots were sold worldwide in 2012 alone that was increased by 12% to 178 132 units in 2013. The International Federation for Robotics estimates that the global population of industrial robots in 2013 was in the range of 1.3 and 1.6 million units with a market value of $9.5 billion. Although most industrial robots are currently found in manufacturing applications, they are becoming more viable for material handling and logistics applications in the future [2].

An associated technology is autonomous control, in which a vehicle or other device has sufficient intelligence to sense its environment and make independent, local decisions. As the complexity of logistics systems continues to increase in the future, autonomous control and distributed intelligence offer a robust and flexible means of control.

The prospect of driverless trucks is a potentially disruptive technology that could offer significant benefit to the logistics industry. Driverless cars already have been licensed in Nevada, Florida, and California. Significant social and technological obstacles remain, but driverless trucks could reduce the need for truck drivers, improve highway safety, and significantly reduce transportation costs.

By 2025, broadband integration of several of these technologies into innovative and coordinated systems could result in revolutionary change in the industry.

6.3.3 *Sensors and C-IoT*

In 2009, Kevin Ashton took credit for the phrase "The Internet of Things" as the title of a presentation made to Proctor & Gamble in 1999. Ashton postulated that humans depend on physical things and value them far more than information because things are what we eat, wear, and use in our daily lives. On the other hand, the Internet deals with data and information, and while information about things can help us improve systems, such data must be put into the Internet by humans who have very limited time and attention.

In 1999, Ashton saw radio frequency identification (RFID) as a mechanism by which physical things could directly communicate with the Internet. "If we had computers that knew everything there was to know about things – using data they gathered without any help from us – we would be able to track and count everything, and greatly reduce waste, loss, and cost. We would know when things needed replacing, repairing, or recalling, and whether they were fresh or past their best."

Since then, the proliferation of embedded sensors that can communicate with the Internet without human intervention is staggering. Consider that today GPS allows real-time tracking of cars and trucks and people through mobile phones. Strain gages rest on structural members of bridges that automatically broadcast critical information to alert highway engineers of potential problems. Vision systems can identify defects in high-speed production environments; so non-complying products can be automatically ejected from the stream prior to shipping. RFID tags attached to shipping containers can record important measurements such as drop forces that the container experiences and a continuous recording of temperature in the container during transit.

Every year, sensor technology is creating smaller and better devices that can "talk" to the Internet without human intervention. The increasing array of functions that these sensors perform is also advancing at an incredible pace, as is the accuracy they can achieve.

A recent McKinsey Quarterly report asserted: "The widespread adoption of the IoTs will take time, but the timeline is advancing, thanks to improvements in underlying technologies. Advances in wireless networking technology and the greater standardization of communications protocols make it possible to collect data from these sensors almost anywhere at any time. Ever-smaller silicon chips for this purpose are gaining new capabilities, while costs, following the pattern of Moore's Law, are falling."

By 2025, sensors that automatically communicate with the Internet without human intervention could be almost ubiquitous. Every step of the manufacturing process could have sensors communicating directly with the Internet, so operators would be warned of problems and be told precisely what to do. End-item packages, unit-load containers, and transportation containers could have continuous GPS tracking-optimizing routing and delivery decisions. Containers should have sensors communicating vital information in real time, such as shock and temperature so remedial actions can be made if an unsafe condition is encountered.

6.3.4 *Big Data and Predictive Analysis*

The term "Big Data" refers to extraordinarily large data sets that companies and other organizations now collect and store about their operations, sales, customers, and nearly any other transaction of interest. How does hurricane activity in the Atlantic or flooding in Texas affect our business? Do customers buy Product A or Product B in Store X? Such information and answer are embedded in Big Data.

The expanding field of data analysis is rooted in classical statistics, but advances in computing power and the availability of massive quantities of data have led to new techniques of data mining and data visualization. Data mining is the science of finding patterns and correlations among (possibly disparate) sets of data. The presence of such patterns can lead to better decisions in logistics and other operations. For example, knowing that customers tend to buy beer and salsa on Fridays during football season can lead grocers to anticipate this demand and be ready to meet it with sufficient stock. New techniques in data visualization allow decision makers to consider large quantities of data very quickly and therefore to make better decisions.

Predictive analytics is a related concept that uses data mining and other techniques to predict the future. It differs from forecasting in that the latter applies mathematical relationships directly to historical data to predict future values (demand, for example), while accounting for variation, trends, and seasonality. Predictive analytics looks for correlation between past, perhaps disparate events, and predicts future events based on current and emerging conditions. For example, the presence of certain terms on social media might portend a shift in demand for fashion items.

By 2025, these techniques will be much more mature, but likely they will not have been fully deployed. Furthermore, exploiting Big Data presupposes that the data are available. Companies are naturally reluctant to share sensitive data that might be of benefit to other parts of the supply chain. The material handling and logistics industry must find ways to make appropriate data available to all who need it, while protecting the interests of owners of that data.

6.3.5 *The Changing Workforce*

Changing demographics in the United States suggest that the challenge in attracting, training, and keeping an adequate workforce for the future will be even greater by the year 2025. The so-called "baby boomers" will be retiring in droves and will be replaced by fewer workers in the next generation.

Attracting and keeping an adequate workforce will require the field to appeal to a much different workforce than it does today. Women, workers under the age of 35, people with disabilities, and veterans are all primary targets to replace the current workforce. To attract these new people, the industry will need to establish a coordinated effort to position material handling and logistics jobs as rewarding careers that are personally fulfilling with many opportunities to advance.

With respect to skills, there is considerable concern for the existing and future workforce. On the one hand, there is a high rate of change in the technologies used and skills required to operate low-cost supply chains every day. On the other hand, skill sets of new employees at all levels are lacking. Beyond technology skills, other gaps include problem-solving abilities, situational response skills, abstract reasoning, and even basic work ethic. These problems will have to be addressed by secondary and vocational-technical schools.

By 2025, the Industry and Government must have in place new initiatives to find, attract, and retain the workers that are necessary for its success – and do so in the presence of many other industries competing for the same talent.

6.3.6 *Sustainability*

Since the financial crisis of 2009, businesses understandably have been focused on surviving the shock and getting lean and agile enough to prosper in turbulent times. Talk of sustainable systems has been overwhelmed in the national dialog by other concerns in society, the economy, and international politics.

Consider the most widely accepted definition for sustainable development, which was given by the World Commission on Environment and Development in 1987 and subsequently endorsed by the United Nations at the Earth Summit in 1992: "Sustainable development is development that meets the needs of the present without compromising the ability of future generations to meet their own needs." The underlying principle hardly seems controversial. Every generation wishes to leave the world a better place than it found it. To believe otherwise is to accept inevitable decline.

A commonly accepted framework that applies this definition to business strategies for private and public organizations identifies three main considerations: economic development, environmental preservation, and social development. In the context of supply chain systems, economic development means the creation of economic value for employees, customers, and stakeholders. Environmental preservation addresses the environmental impact of supply chain operations such as effects on local wildlife, solid waste generation, and emission of pollutants. Social development accounts for the effects of supply chain activities on human populations and societies, including positive effects such as education, and negative effects such as pollution on public health.

By 2025, Industry should have developed standard methods of incorporating sustainable development into business plans and operating strategies. Such methods should adhere to the goals of sustainability, while maintaining and even advancing the commercial interests of the industry.

References

[1] Internetretailer https://www.internetretailer.com/2013/03/13/us-e-commerce-grow-13-2013 (accessed 10 December 2014).

[2] IFR http://www.ifr.org/industrial-robots/statistics/ (accessed 10 December 2014).

7

Conclusion

What exactly is the Internet of Things (IoT) and what does it mean to you? Well, the big picture is pretty straightforward. As described throughout the book, IoT will bring about smartness that will lead to smart living, better quality of life, and higher process efficiency.

Everything will be smart, networked, and automated: Not just devices such as phones and watches, but home and industrial machines and everyday objects such as doors and clothing. They will all have embedded sensors that communicate information to other smart "things" over wireless and wired networks.

And the best part: everything will happen without human intervention. For example, plants and agricultural crops will water themselves when they need it … not when they do not.

7.1 Simple C-IoT Domains and Model

The book has taken a simplified visionary approach for the future of IoT and presented a new model for C-IoT (Collaborative Internet of Things) with focus on collaborative intelligence that will impact our connected life and businesses.

The simplified C-IoT model, in its simplest form, consists of sensing, gateway, and services. Sensing will tap into what matters, and gateway will add intelligence and connectivity for action to be taken at the local level and/or communicate information to a higher level. The services will capture information, digest, analyze, and develop insights of ways that help improve quality of lives or business operation.

Also, the book introduces simplified C-IoT domains consisting of 3Is: Individual, Industry, and Infrastructure where key target business applications can span any or all of these domains. Individual for smart living covers consumer electronics and wearable devices, smart homes, and smart connected cars. Industry for business efficiency covers several markets associated with industry such as factory automation,

Collaborative Internet of Things (C-IoT): For Future Smart Connected Life and Business, First Edition.
Fawzi Behmann and Kwok Wu.

smart buildings, and smart retails. Infrastructure for smart communities and cities for sustainable environment and living includes public transportation and highways, public safety, disaster management, smart education, and smart health care.

7.2 Disruptive Business Applications of C-IoT

Gaining insights in these domains and business applications will result in driving better results. With a C-IoT, smart homes and businesses will make themselves energy efficient and safe.

Examples for some of the domains are discussed in the following sections.

7.2.1 *Individual*

- Smart kitchens will keep a database of supply levels and let you know when you are low on milk and bread or order your groceries automatically. You will even be able to preheat your oven on your way home or have it message you and the kids when dinner is ready. If that sounds too futuristic, I assure you that it is not.
- Smart cars and highways will not only find the fastest way for you to get to where you are going, but also help you get there safely by sensing potential accidents, break failure, and even tire blowouts before they happen.

7.2.2 *Industry*

- Improving energy efficiency with intelligent power grids, self-managed office buildings and industrial plants, and smart meters.
- Using Big Data and software analytics can help predict industrial equipment failure before it happens and reduce downtime.
- Networked smart meters will guide you to free parking spaces as you approach, and your favorite Italian restaurant will tell your car or phone how long you will have to wait for an open table.
- Hospital operating rooms will sense unacceptable levels of bacteria on a surgeon and emergency rooms will be able to locate and route physicians and equipment in real time.
- Smart buildings are leveraging emerging IoT technologies to become even smarter. According to the latest Global Sustainability Perspective from Jones Lang LaSalle (JLL), six advances in smart building technology are enabling a new era in building energy efficiency and carbon footprint reduction, yielding a return on investment for building owners within 1–2 years [1].

The future of C-IoT will manifest itself from two areas: having better product and services and managing our lives more efficiently. These in turn put demands on every aspect of the supply chain for product and services.

If at every stage in the supply chain, the stage suppliers strive to deliver their best, then the sum total of efforts manifested at each stage will have much more positive impact on everyone.

C-IoT helps to tap into resources that have been passive and bring about stream of data that can be collected and analyzed empowering us as to what makes sense out of everything.

Companies need to innovate more and collaborate with other, third parties to deliver a total solution. This will require more innovative approach across the supply chain from SoC, to chip, to module, to board, to equipment, to software, and to applications.

Our lives will transform into a digital lifestyle of better quality, better health, improved fitness all due to the emerging tools and gadgets that bring a new dimension of information at our fingertips, which helps us to make real-time quality choices.

Device lifecycle management will be more important than ever as humans will be truly interacting and subscribing to information from sources that we do not control.

7.3 A New Digital Lifestyle

By the year 2020, connected devices will become the reality we have been waiting for. Our cars will report how much fuel we used; meters will control our thermostats; and pedometers will unlock our refrigerators only when we have walked the required amount of daily steps. Physicians and caretakers will know that something is wrong with our organs before we do. We will be inundated by more ISM (Industrial Scientific and Medical) short-range protocols than we can imagine embedded in our shoes, pacemakers, medicine bottles, wallets, eyeglasses, watches, dog collars, trash bins, and more. Sensors will be embedded in bridges, ceilings, and roads. Bluetooth Low Energy, ANT and mesh networks, ultra-low-power, and short-range wireless technology designed for sensor networks will provide connected "dust" that reports petabytes of information back through gateways requiring petaflops of computing power.

7.4 Development Platform

According to a new survey by market research firm Evans Data, 17% of the world's software developers are already working on IoT projects. Another 23% are planning to start an IoT project within the next 6 months. The most popular devices are Security and surveillance products, connected cars, environmental sensors and smart lights, and other office automation tools [2].

Both Apple and Google have made moves toward building out successful mobile ecosystems – iOS for Apple and Android for Google – into hubs for innovation for IoT fixtures, particularly smart homes. Google acquired Nest Labs, which then opened its Nest Developer Program to make its smart home platform an open-source hub through which third-party manufacturers' devices and outside developers' apps can communicate. Apple's HomeKit, similarly, is being developed as a framework for the communication and control of smart home devices [3].

7.4.1 Influencers for Smart Connected Homes

A recent study report "Internet of Things Influence Study" published by Appinions in July 2014 [4] reveals that the top three influencers were all involved in connected home:

- Apple took the #1 spot by announcing HomeKit, a development platform to integrate existing and future smart home technology into the Apple iOS so that end users can use a single app for all home controls.
- Google's recent acquisition of smart thermostat maker Nest Labs and Nest's subsequent acquisition of security cam firm Dropcam pushed Nest and Google to second and third.

7.4.2 Influencers for Industrial Internet

- GE has coined the term "industrial internet" to unite the fields of Machine-to-Machine (M2M) communication, manufacture, big data, and the IoT and led the way in establishing the Industrial Internet Consortium to bring together firms interested in developing this type of technology.
- AT&T and IBM partner on IoT for smart cities and utilities. The companies will combine their analytic platforms, cloud, and security technologies with a focus on privacy. AT&T will manage sensors' communications and tracking happening over the cellular network, and IBM will use its analytics platforms. They are targeting customers and using cases that produce vast troves of data from devices as diverse as mass transit vehicles, utility meters, and video cameras. The goal is to use their resources to identify patterns and trends to improve urban planning and let utilities better manage their equipment to reduce costs [5].

7.5 C-IoT Emerging Standards, Consortiums, and Other Initiatives

Over the past several years, there were many books, articles, seminars, and events held on IoT. Today, IoT is an integral part of many of the conferences. Many companies have allocated a business unit to address IoT markets, common platform, interoperability and testing, and other areas of concerns. As a result, many events such as seminars/webinars and workshops were conducted in different regions of the world to raise awareness and provide input to other forums. Consortiums were created in 2014 and several international standard organizations have started to work on developing standards for IoT.

The following section provides some of these initiatives with the hope of encouraging the reader to get engaged and help advance the progress of IoT standards that will help in rapid prototyping, and development and deployment of innovative applications and solutions.

7.5.1 C-IoT Emerging Standards

Most of the following list of standards activities was created in 2013/2014.

7.5.1.1 IEEE – Standard for an Architectural Framework for the Internet of Things (IoT)

IEEE Standards Association on Innovation and IoT is currently working with over 20 companies in developing architectural standards for IoT under Project P2413. The IEEE-SA Board of Governors/Corporate Advisory Group sponsors the Project.

This standard defines an architectural framework for the IoT, including descriptions of various IoT domains, definitions of IoT domain abstractions, and identification of commonalities between different IoT domains.

The architectural framework for IoT provides a reference model that defines relationships among various IoT verticals (e.g., transportation and healthcare) and common architecture elements. It also provides a blueprint for data abstraction and the quality "quadruple" trust that includes protection, security, privacy, and safety. Furthermore, this standard provides a reference architecture that builds upon the reference model. The reference architecture covers the definition of basic architectural building blocks (Sensing, Gateway, and Services in this book) and their ability to be integrated into multi-tiered systems solution [6].

7.5.1.2 IEC – Smart Grid

The IEC is a not-for-profit organization that brings together 165 countries and offers a global platform to over 13 000 experts from industry, governments, and user groups. These experts define specifications, measurement methodologies, and testing requirements that are needed to do business in the global market. They develop International Standards that cover all aspects of safety, interoperability, efficiency, electromagnetic compatibility, and environmental impact. One of the target businesses is Smart Grid, which includes Smart Grid Standards Map, IEC standards, strategy, interoperability, regulations, and others.

Electric energy is the ultimate just-in-time product. It needs to be used the moment it is generated and must be supplied continuously. Today's Grids are deeply rooted in technology that was modern more than 100 years ago, long before the first microchip. Most Smart Grid Project managers are now charged with updating those legacy systems. And the big question is how [7].

7.5.1.3 ITU and IoT

International Telecommunication Union (ITU) has established several groups 2, 3, 9, 11, 13, 16, and 17 as well as Joint Coordination Activity on IoT and focused group on M2M service layer in advancing the work on IoT [8]:

- *JCA-IoT*: Joint Coordination Activity on Internet of Things
- *FG M2M*: Focus Group on Machine-to-Machine Service layer
- *Study Group 2*: Operational aspects of service provision and telecommunications management
- *Study Group 3*: Economic and policy issues
- *Study Group 9*: Broadband cable and TV
- *Study Group 11*: Signaling requirements, protocols, and test specifications
- *Study Group 13*: Future networks including mobile and NGN (next-generation network)
- *Study Group 16*: Multimedia coding, systems, and applications
- *Study Group 17*: Security

7.5.1.4 ITU and Intelligent Transport Systems

There are ITU efforts focusing in development standards for Intelligent Transport Systems (ITS). Focus is to improve traffic flow, to increase the efficiency of freight and public transportation, and to reduce fuel consumption. They also have been identified as a tool to improve road safety. This report analyzes the functionalities of ITS and describes the set of technologies used in ITS. It reviews current ITS standardization activities and identifies possible areas for future ITU-T work [9].

7.5.1.5 ISO – Intelligent Transportation Systems, e-Health on IoT Matters

To prepare the future of road vehicles, ISO works closely with its partners of the World Standards Cooperation (WSC) – the International Electrotechnical Commission (IEC) and the ITU [10].

7.5.1.6 LTE for Cellular IoT

A group of industry players, including network and device suppliers, operators, and academics, brought together by Vodafone, has been looking at the problem of supporting the "IoT" for the past year and has recently published a White Paper outlining the options. These include further new features in LTE (Long-Term Evolution) that would be defined through 3GPP; or an alternative "clean slate" cellular standard defined specifically to meet the needs of the IoT. Any future system will need to connect "things" that do not have large amounts of data to communicate; can be in hard-to-reach locations such as manholes, meter closets, and in very isolated locations; and need to operate for years on small batteries [11].

7.5.2 *C-IoT Emerging Consortiums*

7.5.2.1 Internet of Things Consortium

It was formed to drive adoption of IoT products and services through curated networking, consumer research, and Industry Education. The value proposition is to make consumers' lives more efficient, safer, and seamless. Member companies include SmartThings, Konrol TV, OUYA, Logitech, Plum, LUMBO, Planet Labs, Mojio, Flatout, Koubachi, iotlist, techstars, O'reilly, and Solid [12].

7.5.2.2 Open Interconnect Consortium (OIC)

It is an IoT consortium that is aimed to set standards for connecting billions of household gadgets and appliances. The Open Interconnect Consortium (OIC) is focused on delivering a specification, an open source implementation, and a certification program for wirelessly connecting devices. Member companies include Atmel, Broadcom, Dell, Intel, Samsung, and Wind River [13].

7.5.2.3 AllSeen Alliance

The organizations involved in AllSeen work off of Qualcomm's AllJoyn open source project as the initial framework. Members include Qualcomm, LG, Electronics, Sharp, Panasonic, Haier, Technicolor, Silicon Image, TP-LINK, and others.

7.5.2.4 Industrial Internet Consortium (IIC)

The Industrial Internet Consortium (IIC) is an open membership organization formed to accelerate the development, adoption, and widespread use of interconnected machines and devices, intelligent analytics, and people at work. IIC has over 50 members. Its Founding members include AT&T, Cisco, GE, Intel, and IBM [14].

7.5.2.5 Thread

Thread was designed with one goal in mind: to create the very best way to connect and control products in the home. Founding members include ARM, BIGASS, Freescale, Nest, Samsung, Silicon Labs, and Yale [15].

7.5.3 Forums, Workshops, and Other Initiatives

7.5.3.1 Worldwide IoT Standards Workshops

- 2012: Beijing, China and Milan, Italy
- 2013: Shenzhen, China and Mountain View, CA
- 2014: IEEE IoT World Forum, Seoul, Korea (March 6–8)
- 2014: IoT Developers Conference, Santa Clara, CA (May 8)
- 2014: IoT Week, London, UK (June 16–20)
- 2014: IoT World, Palo Alto, CA (June 18)
- 2014: Workshop, Mountain View, CA (September 18–19)
- 2014: The IoT World Forum, Chicago, IL (October 14–16)
- 2014: Healthcare 2014 Workshop – E-health, IoT, and Cloud: standards, challenges, and opportunities, Natal, RN Brazil (October 15–18)
- 2014: IoT Applications USA, Santa Clara, CA (November 19–20)
- 2014: IoT World Forum, London, UK (November 25–26)
- 2015: Third International Workshop on Pervasive IoT and Smart Cities, Gwangju, South Korea (March 24–27)
- 2015: International Conference on Recent Advances in IoT, Singapore (April 7–9)
- 2015: IoT North America, Chicago, IL (April 15–16)
- 2015: IoT Applications Europe, Berlin, Germany (April 28–29)
- 2015: IoT Developers Conference, Santa Clara, CA (May 6–7)
- 2015: Second Annual IoT World, San Francisco, CA (May 12–13)
- 2015: Internet of Things Expo, New York, NY (June 9–11)
- 2015: Internet of Things Expo, Santa Clara, CA (November 3–5)

7.5.3.2 IoT Industry Roundtables and Webinars

- 2012: Milan
- 2013: Korea and the United States
- 2014: Four Industry Roundtables in the United States, Europe, and Asia
- 2014: IoT Day, Global via Meetup (April 9)
- 2014: IoT How Are European and US Regulators Responding? (March 24)

7.5.4 C-IoT and Radio Communications

The integration of short-range ISM radios such as Wi-Fi, Bluetooth, ZigBee, and Z-wave on modules will proliferate. Radios will become more application specific with software embedded for wireless and wired protocols such as Modbus over Ethernet and ZigBee over 802.15.4. I also believe that Java will abandon Linux and will go bare metal in order to improve on resources, energy, and startup time.

7.5.5 C-IoT and Nanotechnology

Nanotechnology will play important part in the future C-IoT. Some examples were presented in the future section at the beginning of the book. Fundamentally, nanotechnology will converge with Semiconductor and is expected to revolutionize every industry. One of the impacts is on the life and size of batteries.

7.5.5.1 Nanotechnology and Semiconductor

According to Dr. Eric Drexter, University of Oxford, a pioneering nanotechnology researcher and author, we are likely to see transition to a new industrial revolution and a large change in manufacturing. Convergence between nanotechnology and semiconductor technology can take digital information systems beyond the limits of Moore's law. Semiconductor industry is well positioned to benefit from advances in nanotechnology.

7.5.5.2 Nanotechnology and Batteries

Nanobatteries have increased surface areas because the electrodes are coated with nanoparticles.

Batteries using nanomaterial can increase the life of batteries, reduce the size, and release no current when it is not in use.

Current silicon when switched off, it still consumes current. But with nanotechnology new material and transistor, there will not be any leakage. Semiconductor (with 15 nm and beyond) will capitalize on nanotechnology to reach 1 TB speed. As you reduce the size of the silicon (2000 times smaller than that of the diameter of your hair), you can pack more processing power and other components.

Innovation will take place by using nanotechnology with Nanophotonics (light – current global market size is $1.8B – Transparency Market Research) to be used to transmit the data directly on the chip, which has higher speed and uses less power in transmitting data.

7.5.6 C-IoT and Security

The transient effects of large numbers of connected devices and the coordination of those devices in situations of anomaly will become more important as the number of devices will outnumber handsets 10-fold. Some of these devices (e.g., person–company–company, person–vehicle–vehicle, person–government, and Robot–Robot) will have both publishing and subscription capabilities, communicating on a peer-to-peer basis. As these devices are deployed, the complex interaction between them and the network will cause perturbations that could have a detrimental effect on the network, much like "packet storms" in IP networks.

This pending social network of machines has implications for security, too. If we subscribe to other companies, the government, or even a sensor crowd source of information, how do we know that it is secure and accurate? Security will be the next important leap forward in IoT.

Apple and Google are two of the biggest players in the IoT market. Google Glass provides fast access to information by speaking commands to the microphone built into the smart eyewear device. In 2014, Google acquired smart thermostat company Nest for $3.2 billion. Nest is best known for creating the Nest Learning Thermostat, which learns the temperature preferences of its users. Smart thermostats are expected to be a big market in the next few years – $1.4 billion by 2020, according to Navigant Research [16]. Google also acquired a Wi-Fi-enabled security camera company Dropcam for $555 million. Recently, at WWDC (Worldwide Developers Conference), Apple announced a new smart home framework called HomeKit, which can be used for controlling connected devices inside of a user's home.

7.6 Final Note

With growing adoption of smart phones and social media, citizens or human-in-the-loop sensing and processing user-generated data and data generated by user-wearable/mobile devices continue to be key sources of data and information about us and the markets around us. Better insights will be gained through cognitive computation coupled with business intelligence and visual analytics that is GIS (geographic information system) based. This helps in improving judgment and speed process of taking a better decision.

This unified smart C-IoT software platform enables one to build and deploy smart C-IoT product, systems solutions, and services for different vertical markets in a quick time-to-market fashion. The C-IoT service platforms leverage sensor fusion software framework components, to perform ubiquitous sensing and processing works transparent from the user. The will enable the development of C-IoT applications that interoperate among multiple point solutions of Smart things. This will further drive the IoT market products to new heights.

Billions of devices and sensors, all communicating through the cloud and all feeding into a massive analytics solution to provide a complete picture of the business, its employees, its processes, and its customers is exactly how these technologies should be coming together.

This book encourages IoT Global collaborative innovation where people from diverse background and talent can contribute with ideas, research for advancement of technology, disruptive approaches to applications, and services.

References

[1] M2M http://m2mworldnews.com/2013/06/27/45768-m2m-technology-convergence-making-smart-buildings-even-smarter/#sthash.1axML03a.dpuf (accessed 18 November 2014).

[2] CheatSheet http://wallstcheatsheet.com/technology/can-google-apple-or-samsung-lead-the-internet-of-things.html/?a=viewall (accessed 18 November 2014).
[3] Appinions http://dj.appinions.com/iot-july-2014/ (accessed 18 November 2014).
[4] Fiercewireless http://www.fiercewireless.com/story/att-ibm-partner-internet-things-smart-cities-utilities/2014-02-18 (accessed 18 November 2014).
[5] IEEE Standards Association http://standards.ieee.org/innovate/iot/stds.html (accessed 18 November 2014).
[6] IEC http://www.iec.ch/smartgrid/?ref=extfooter (accessed 18 November 2014).
[7] ITU http://www.itu.int/en/ITU-T/gsi/iot/Pages/default.aspx (accessed 18 November 2014).
[8] Scribd Inc. http://www.scribd.com/doc/17521708/Standardization-Activities-for-Intelligent-Transport-Systems (accessed 18 November 2014).
[9] ISO www.iso.org (accessed 18 November 2014).
[10] 4g-portal.com http://4g-portal.com/lte-evolution-for-cellular-internet-of-things (accessed 18 November 2014).
[11] Internet of Things Consortium http://iofthings.org/#members (accessed 18 November 2014).
[12] Open Interconnect http://www.openinterconnect.org/ (accessed 18 November 2014).
[13] Industrial Internet Consortium http://www.iiconsortium.org/index.htm (accessed 18 November 2014).
[14] Threadgroup www.threadgroup.org (accessed 18 November 2014).
[15] Forbes http://www.forbes.com/sites/aarontilley/2014/01/13/google-acquires-nest-for-3-2-billion/ (accessed 18 November 2014).
[16] WIRED http://www.wired.com/2014/07/platform-wars (accessed 18 November 2014).

Index

Page numbers in *italics* refer to Figures; those in **bold** to Tables

Collaborative Internet of Things (C-IoT): For Future Smart Connected Life and Business, First Edition.
Fawzi Behmann and Kwok Wu.

Printed in the United States
by Bookmasters

Printed in the United States
By Bookmasters